Simplifying Complexity: Life is Uncertain, Unfair and Unequal

Bruce J. West

US Army Research Office

Research Triangle Park

Durham

NC 27709

USA

Simplifying Complexity: Life is Uncertain, Unfair and Unequal

Authors: Bruce J. West

ISBN (eBook): 978-1-68108-217-2

ISBN (Print): 978-1-68108-218-9   © 2016, Bentham eBooks imprint.

Published by Bentham Science Publishers – Sharjah, UAE.

First published in 2016.

Acknowledgements:

The author wishes to thank the Army Research office for supporting the research on which this essay is based and Sharon for her love, understanding and infinite patience.

advertisements or ideas contained in the Work.

Limitation of Liability:

In no event will Bentham Science Publishers, its staff, editors and/or authors, be liable for any damages, including, without limitation, special, incidental and/or consequential damages and/or damages for lost data and/or profits arising out of (whether directly or indirectly) the use or inability to use the Work. The entire liability of Bentham Science Publishers shall be limited to the amount actually paid by you for the Work.

General:

1. Any dispute or claim arising out of or in connection with this License Agreement or the Work (including non-contractual disputes or claims) will be governed by and construed in accordance with the laws of the U.A.E. as applied in the Emirate of Dubai. Each party agrees that the courts of the Emirate of Dubai shall have exclusive jurisdiction to settle any dispute or claim arising out of or in connection with this License Agreement or the Work (including non-contractual disputes or claims).
2. Your rights under this License Agreement will automatically terminate without notice and without the need for a court order if at any point you breach any terms of this License Agreement. In no event will any delay or failure by Bentham Science Publishers in enforcing your compliance with this License Agreement constitute a waiver of any of its rights.
3. You acknowledge that you have read this License Agreement, and agree to be bound by its terms and conditions. To the extent that any other terms and conditions presented on any website of Bentham Science Publishers conflict with, or are inconsistent with, the terms and conditions set out in this License Agreement, you acknowledge that the terms and conditions set out in this License Agreement shall prevail.

Bentham Science Publishers Ltd.
Executive Suite Y - 2
PO Box 7917, Saif Zone
Sharjah, U.A.E.
Email: subscriptions@benthamscience.org

BENTHAM SCIENCE

CONTENTS

FOREWORD

We in Western culture like to believe the world, especially its human corner, is a reason-driven place. We perceive the universe to operate in accordance with fixed laws. For most of us, science has caught up to, if not replaced, religion as a primary mover. The urge to seek and excavate nature's operating rules dates back more than a millennia, inspired by Aristotle and codified into the scientific method by Roger Bacon. People trust in the scientific method partly because we trust scientists to bracket off their initial biases so that they may engage in a uniform system of inquiry.

But what if scientists are unable to bracket off initial biases? What if cognitive biases about the world are actually built into the way science is conducted? If so, then empirical results are not wholly objective, but instead are partially rigged from the start. And what if the population at large unwittingly shares in these biases, often without realizing it?

This is precisely the state of affairs Bruce West reveals in his ground breaking book, Simplifying Complexity. In order to test hypotheses empirically, all research, no matter what the discipline, is shaped and even driven by the framework of its underlying mathematics. As West suggests, scientists' choice of mathematics goes hand-in-hand with their underlying "cognitive maps" that color reality according to various unconscious slants. These cognitive maps are necessary, for their facts and presumptions allow us to simplify the incredible complexity that surrounds us, so that we may function with purpose and free from overwhelm.

For centuries, Western science has relied primarily on linear statistics, whose "normal" bell-shaped curve and regression to the mean yield strong central tendencies. The normal distribution characteristic of linear statistics operates by collapsing variability to a point in the middle, which is then used to characterize the whole. In conjunction with Newtonian mechanics, linear statistics paint nature and her contents in clockwork fashion. Just like the mechanism of a clock can be disassembled and reassembled without surprise, under the influence of linear statistics, so too does Newton and Laplace's world move predictably in small, additive steps. The resulting statistics give the universe the appearance of being both fair and equal. In general, this choice of mathematics serves to tame nature's wild, uncertain, unpredictable, and unjust side.

West likens the mathematics a scientist selects as the formal medium for experimentation to an artist's choice of medium for self-expression. In both science and art, the medium conveys the message, as Marshall McLuhan would claim. Yet, the clockwork picture supported by linear mathematics is only one, highly simplified view of the world—one which easily leads to confusion. As a child raised in the 1960s, I recall learning an important statistic: the "norm" for the number of children growing up in families in the United States was said to be 2.5. Being rather literal-minded, I found it hard to wrap my mind around this notion. What exactly does it mean to have 2.5 children? When it comes to real families, it seemed to me that no one had 2.5 children. Looking back, I can see that wrestling with these ideas presaged my current interest in nonlinear science.

As West explains, we easily take for granted the story created by our mathematics of choice, largely due to familiarity. Normal statistics work well for sampling height or weight in the general population, because variability among people is relatively small; differences are additive; and underlying elements remain independent from one another. Yet, these conditions do not apply to most complex systems as they operate in the real world. Most natural distributions show wide spreads that include extreme if not catastrophic events. Indeed, catastrophic events occur far more often than most of us would like to believe. In general, nonlinear statistics reveal nature's wild side—her uncertain jolts, abrupt transitions, unpredictable turns, and dynamic variability in general.

With respect to real people in the real world, life is not clockwork nor is it mechanical, and families are not normative. Some people have 1 child; others have 10. When taking into consideration generation after generation that precedes the current one, each family looks different. The more families we sample, the greater the variability we find. Life thrives on variability. And the faster contemporary society changes with the advent of new technology and communication devices, the more dynamic the surrounding variability. Through nonlinear lenses, variability is the new norm, and for this reason, the time is right to shift our cognitive maps.

In this book, West presents a different kind of metric with which to understand nature's complexity and refine our underlying assumptions about the world. He offers nonlinear statistics that capture variability in the distributions of objects in space as well as events in time. In contrast with linear statistics that apply primarily to simple systems whose constituents are independently organized, nonlinear statistics apply to complex systems whose interdependent elements shift in multiplicative or exponential ways. This type of statistic goes by several names—1/f distributions, inverse power laws, pink noise, Pareto's law. West outlines as many as 9 different mechanisms giving rise to similar surface distributions.

One consequence of variability at the heart of such distributions is the importance that extreme events take. This is consistent with how the human mind works: we hardly take notice of tiny fluctuations in ordinary life, but do pay attention to extremes. Every time I successfully pull my car out of the garage, it is a non-event; meanwhile, the time I smash the side of my car is the single instance that matters most. Society at large works the same way.

Consider the stock market: minor fluctuations hardly make the news, but a market crash carries reverberations for years, if not decades, to come.

We human beings are complex creatures. Our brains have more interconnections among their neurons than the entire number of atoms in the universe. Our brains teeter on the edge of chaos, displaying some amount of order, yet enough variability for quick adaptation to an ever-changing environment. As West demonstrates, there is even variability at the center of our beating hearts—quite literally. Whereas traditional Western medicine asserts health in the form of predictable stability and regularity, when examined at the micro level, the dynamics of a beating heart reveal quite the opposite state of affairs. In between each and every heart beat is a tiny bit of variability that keeps us resilient and healthy.

This kind of variability is the stuff of life. As a clinical psychologist, I am steeped in it. I have probably seen hundreds of depressed people in the course of my 30+ year career. While linear statistics might put them all in the same diagnostic box, to me in real life, no two have ever looked exactly alike. Indeed, if they did, I would have been bored out of my mind and not skilled enough to treat them. The closer I look at each person—whether depressed or not—the more unique that individual appears. Welcome to the realm of the nonlinear.

As you turn the pages ahead, prepare to go through Alice's rabbit hole. For once you understand the characteristics of nonlinear statistical distributions, some ordinary assumptions about certainty, fairness, and equality in the world will be turned upside down. When it comes to being complex systems living in a highly complex, interdependent world, dynamics are ever-changing and the future is uncertain. We may use tricks like distraction or mindful awareness to tolerate the anxieties and preoccupations that living with ambiguities and uncertainties entails. Meanwhile, the democratic ideal of everyone having a fair and equal shot at becoming rich, famous, or President, is statistically opposite to the real state of affairs. The rich keep getting richer; the poor keep getting poorer; those already famous keep getting more famous; while scientists whose work is most cited will keep getting more credit, even if they had little to no hand in the underlying research.

From my perspective as a psychotherapist, the linear version of the world appears to be a nice fairy-tale people paint to ease the scariness, unfairness, and inherent injustice of life. To me, this cognitive map of the universe is akin to how parents appear to young children—bringing up the sun and moon and in control over everything. The illusion of certainty and control yields a safe and predictable world that protects little ones from harm and worry. Perhaps this naïve story of our environment, both inside and outside, serves a similar purpose within the developmental trajectory of science. Maybe during the early phases of scientific exploration, scientists similarly needed to keep things simple, by warding off the highly complex and unpredictable side of life.

West's book teaches us an important lesson: it is time for humankind to grow up and wipe the fairy dust from our metaphorical eyes. Simplifying Complexity reveals fascinating facts that connect separate disciplines with the same underlying mathematics. West even introduces a candidate for the first universal principle to govern the interaction of complex systems. West's complexity management cube applies to diverse phenomena—from habituation in a brain that encounters a strong smell to the de-habituation in a brain that becomes riveted on a beautiful piece of classical music. His new principle even addresses issues related to modern warfare and global warming, at points to yield surprising, if not controversial, results.

Most of us have been so thoroughly steeped in linear, reductionist assumptions about how the world works, we act like fish happily swimming in the calm waters of our protected bowls, oblivious to the turbulent waters in the real world outside. But we can't live in isolated environments (including ivory towers) forever. Just as it is dangerous to remain a child sheathed in the false comfort of a predictable, controllable, and fair future, so too is it dangerous to remain naïve about nature's implicit inequities and injustices. By understanding nature's true complexity, as rendered transparent by West (impressively without the use of a single mathematical equation), we prepare ourselves to address the complex problems we each face, both individually and collectively.

Terry Marks-Tarlow, Ph.D.
Santa Monica, CA
USA

PREFACE

The present book is an extensive revision of the previously published "Complex Worlds, Uncertain, Unequal and Unfair" and the title was changed to reflect that change. I believe an ebook publication will increase the likelihood of reaching an audience that is curious about what science can offer the first generation born into a mature information age.

A general polishing of the presentation has been made throughout the revision, but the most significant changes involve incorporating suggestions made by readers. One such change is the emphasis on using the powerful methodology of network science to guide the making of individual and corporate decisions in our complex society. There is also additional discussion on how a new way of thinking is required to fully utilize the results coming out of the new intellectual maps of the complex world of the 21st century.

Bruce J. West
Research Triangle Park
Durham
NC 27709
USA

CONFLICT OF INTEREST

The author confirms that author has no conflict of interest to declare for this publication.

ACKNOWLEDGEMENTS

The author wishes to thank the Army Research Office for supporting the research on which this essay is based and Sharon for her love, understanding and infinite patience.

PROLOGUE

One of our strongest urges as human beings is to know the future in order to control our destiny and the destiny of those in our charge. Therefore, proposing the notion that life is uncertain, unequal and unfair seems to undercut a basic human need. But my intention in writing this book is not to subvert that primal need, but just the opposite. My belief is that the more clearly a person understands how the world actually works, the more effective they can be in achieving what they want in life. My experience is that individuals are ineffective, in large part, because they fail to distinguish between how the world is and how they want it to be.

I am not a psychologist, sociologist or social worker. I am a physicist and as such I am more comfortable with mathematical equations and experimental data than I am with words. So this is not a self-help book. It is a somewhat personalized account of what the average college graduate in western society has been systematically taught about the scientific understanding of the world that is not true. The greatest myth that has been delivered with all the pomp and circumstance of scientific truth is that for most purposes a linear view of the world is more than adequate. This is where the book begins, tracking down some of the less obvious implications of a linear world view and how that view conflicts with available data.

I heard somewhere that an editor told a would-be author of popular science books that with each equation she would lose half her intended audience. I do not know if this is true, but to be on the safe side I have not included any equations in the book, but there are charts and graphs along with ample interpretive discussion to replace them. This is a book about science and the role science plays in our technological society, both on stage and behind the scenes.

A brief review of the historical evidence that the mental maps of the world we construct for ourselves consist primarily of elements that are linearly connected provides a context to understanding why some people refuse to act in their own self-interest. Even when uncertainty is introduced into the description of events, as a way of including the influence of the broader world into their development, that uncertainty takes the form of small, additive, random fluctuations. The world's ambiguity is represented by a bell-shaped distribution of fluctuations in the outcomes of experiments and the variability of observations. The bell-shaped distribution reveals certain general properties of the world's influence on simple predictions, whether it is your broker's estimate of the likelihood of a market crash, your doctor's estimate of the severity of your disease, or whether you should have received a raise rather than the new guy.

I examine how the neatly constructed linear world view has been challenged by the complexity of modern society. It is not the case that humans have changed how they construct their mental maps of the world. It is that the linear assumptions made in the past are no longer as useful, as they once were, in guiding decisions made, particularly when social interactions are long range, multiple, and anonymous. I will indicate how the disintegration of simplicity disrupts our lives and leads to such things as the mismanagement of the health care system, particularly through the dominance of extreme events, when the assumption of Normal statistics no longer suppresses outliers. Specifically I am concerned with the form in which the notions of fairness and equity, born in the social unrest and industrialization of the nineteenth and early twentieth centuries, survive in the data of the twenty-first century; or more accurately how they do not survive or have been transformed.

Much of the fear that is generated by those that misapply the notion of complexity is based on extrapolating recent fluctuations into the future. Such extrapolation is invariably done using linear models that almost never have anything to do with the phenomenon being extrapolated. This was done using the "science" of eugenics at the turn of the twentieth century, and was the scientific basis for the "Aryan race" so loved by Hitler and still considered fondly by "white supremacists" and "skin heads" today. A similar kind of scientific basis was made in the 60s and 70s for overpopulation and "global winter". These things are mentioned in passing, but what is important is that we must abandon the idea that complex phenomena lend themselves to simple linear predictions.

Such a strong conclusion requires an abundance of evidence regarding the value of replacing linearity with complexity. The implications of a complex representation of the world are immediate and profound. One inherent advantage is that the complex vantage point provides a single coherent view of disruptive mechanisms in complex phenomena; mechanisms ranging in physical science from earthquakes to floods; in social science from stock market crashes to the failure of power grids; in medical science from heart attacks to flash crashes in health care; and in biological science from the extinction of species to allometry relations. Extrema are more frequent in the complex world than they are in the linear world. The effects of extreme events are certainly unfair, and fortunately they do not occur every day. But when disruptive events do occur they introduce crossroads, and the selection of which road to take determines the subsequent course of events in a person's life. Consequently, understanding the source of extremes enables an individual to take back control from the hands of fate.

The transition of our mental models from a simple to a complex world view, entails the breakdown of bell shaped statistics and necessitates the adoption of inverse power-law distributions. This is nowhere more evident than in the distribution of wealth. The long tail in the inverse power law implies that there is a fundamental imbalance in how wealth is distributed and this imbalance was identified by Pareto, the engineer that first identified the effect over a century ago. We shall explore whether or not such imbalance is necessary in a stable social society. This is done by studying other, less emotionally charged, phenomena that share many of its properties. To compare physical, social and biological systems it is necessary to have a common language and for this the idea of an information-dominated system is introduced and developed. The appropriate quantities to measure in complex dynamical systems are not easy to identify, in fact, what we choose to measure may well be determined by how we define information and how that information changes in time. How information flows in complex networks, or how information moves back and forth between two or more complex networks, is of fundamental importance in understanding how such networks or networks-of-networks operate. This information variability is shown to be determined by inverse power-law distributions, which in turn are generated by a number of generic mechanisms that couple contributing scales together. We identify different mechanisms that produce empirically observed variability; each one prescribing how the scales in the underlying process are interrelated.

Science is about finding order in the panorama of the world and embracing a perspective that includes the falling of apples and the motion of planets; the behavior of the individual and the actions of groups, large and small; the information content of an encyclopedia and the wikipedia; in short, science does not, and should not, have any boundaries with regard to content. The terrestrial and the cosmic are part of the give and take in science, with the goal of uncovering the principles and laws that determine how the universe functions, along with the individuals within it. For most people, science appears to be separate and apart from the world in which they live. The principles and laws of science do not seem to apply to the general interactions among people; due, in part, to the fact that principles have not been found for everyday decision making; laws have been notoriously absent from mundane thinking; rules have been sought in vain in the growth of society; and indeed canons go begging in the multiple complex phenomena within the human sciences, despite over two hundred years of effort to either invent or find them. A possible exception to this pessimistic summary of history is given by the Principle of Complexity Management, whereby a system with greater information, but perhaps lesser energy, can dominate a system with lesser information, but greater energy. The principle is a recently proven generalization of an observation made by the mathematician Norbert Wiener, and may be one of these long sought universal principles.

The final chapter contains my understanding of the formal justification for complexity in the real world. In turn, it is an examination of what complexity implies, about the difference between how we react to what we have, as opposed to reacting to what we want, but do not have. People always respond to events according to their mental maps of the world. Consequently, when they find the response to be inappropriate, the most reasonable thing to do is change the map. However, people are not always reasonable or logical. My hope is that the potential for understanding presented in this book can initiate the wisdom that St. Francis addresses in his brief prayer:

God, grant me the serenity
To accept the things I cannot change,
Courage to change the things that I can,
And the wisdom to know the difference

At the suggestion of an anonymous reviewer to present additional discussion on the interpretive strength of nonlinear models I have elected to include an epilogue.

Simplifying Complexity:
Life is Uncertain, Unfair and Unequal

CHAPTER 1

How Scientists Think

Abstract: We begin by focusing on the ways we record the myriad of events that make up our lives, using simple models that are intended to capture the dominant features of those events and to provide coherent interlinking of events. If the world did not change in time, more and more detail could be added to these models, with each repetition of an event. Eventually we would have an accurate reconstruction of a successful economic relationship, of a nurturing family, or of a supportive organization. But things do change, even if our reactions to them do not. To understand these changes scientists have developed techniques that quantify and communicate objective models of these subjective events. Without presenting the technical details of how scientists construct such models, I use a combination of personal history and discussions of the science hidden by a variety of social problems, to lay the foundation for the understanding and resolution of these problems in subsequent chapters.

Keywords: Chaos, Exponential growth, Grand visions, Mental models, Multiple saturations, Saturation, Technology evolution.

Science, as well as the typical scientist, has changed along with society. From its slow paced agrarian roots, to the faster paced industrial form, to the nearly instantaneous informational society, the concerns of science and scientists have steadily expanded. The basic science describing the mechanical motion of the planets orbiting the sun, matched the relatively simple social forms that were directly supported by farms and farmers. The increased complication of the statistical description of the interaction of large numbers of particles in a gas was more compatible with industrial mores and the networks necessary to support them. Finally, tipping points and global interdependence spawned the analysis of complex phenomena in harmony with the information society. This historical tagging of the concerns of science indicates that we tend to think of these distinct social modes as being separated by large intervals of time, say centuries. Although historically accurate, such a picture distorts the influence that these distinct social modes have on individual scientists. So it is not without value to include some personal history in my presentations, since all three social modes have influenced my own development as a scientist.

So I begin with my father, who was brought up on a five hundred acre farm in upstate New York; the oldest of thirteen brothers and sisters. The farm was without benefit of electricity or indoor plumbing, except for a hand pump providing water with which to wash and cook. He graduated eighth grade when he was 12, but there was no high school in his rural community, so he had a choice to either leave school and work the fields with his step father, or stay in school and read every book in the library including two encyclopedias. He choose the latter. Like many young men his vision of the future did not coincide with that of his parents. The world beckoned to him and he left the farm when he was 16; it was the depth of the depression.

My mother was raised on an even smaller farm in upstate New York; the oldest of seven brothers and sisters. Her town did have the luxury of electricity, as well as, a high school from which she graduated. The daughter of Italian immigrants, she was the first in her family to receive a high school diploma. I have one book she kept from that time, *The Logic of Epistemology*, not the kind of high school reading seen today.

My parents were married when my mother was 21 and my father was a few years older. They gave birth to seven boys, three before the Second World War and four after my father came home after serving in the Army on an island in the Pacific. The ages of my brothers were spread over seventeen years; the youngest was in a crib in my room when I left home at 17. I shared my room with four younger brothers. Like my father I was restless and did not share my parents view of the future. I was the third oldest of seven sons, born into a labor class family, and this circumstance contributed significantly to my decision to be a scientist.

My first memory of wishing to be a scientist is associated with a eulogy I wrote on Albert Einstein for an eighth grade English assignment. Thinking about it now

I can see how the idea must have been swirling around in my head for some time, but it took the death of this great man to focus the desire. It was 1955 and once a month there were school drills in which students were guided to duck under their desks in response to an imagined, but no less real, bright flash of light in the sky. We were periodically shown films of cities being destroyed by atom bombs and every Catholic mass ended with the phrase "Savior of the world, save Russia". At that age the 'how' of things seemed much more important than the 'why'. It is only after years of study that I began to understand the reasons underlying the 'why' and to appreciate their entanglement with the 'how'.

Modern science, or more precisely physics, began with Sir Isaac Newton (1642-1727), who famously wrote in response to critics who wanted him to 'explain' the causes of gravity, that he constructed no hypotheses. Newton believed that what could not be directly inferred from experiment constituted hypothesis and he was having none of it. A hypothesis is a refined version of the vague impressions, half-backed ideas, ill-conceived assumptions and intuition that are often generated during the scientific investigation and solution of complex problems. The hypothesis summaries what is learned in the feverish attempt to understand a mystery, but only after the fever has subsided. Scientists typically formulate a hypothesis near the end of a study to make clear to others exactly what it was they were attempting to prove, but only after they are pretty sure they know the answer.

Only extremely simple problems have solutions that can be put into the form of a hypothesis before any research has been done. So when I refer to how scientists think it is not about the formation and testing of hypotheses, but it is about how we acquire knowledge from experiment. What a scientist works to avoid in this acquisition of knowledge is confirmation bias. Such bias was identified by the mathematician/philosopher B. Russel:

> "If a man is offered a fact which goes against his instincts, he will scrutinize it closely, and unless the evidence is overwhelming he will refuse to believe it. If, on the other hand, he is offered something which affords a reason for acting in accordance to his instincts, he will accept it even on the slightest evidence."

Of course everyone is guilty of confirmation bias. I am, you are and everyone I know more readily accepts arguments that supports what they already believe to be true, than arguments that do not. As a private individual such bias determines the books I read, the movies I watch and the friends I have. However as a scientist I must proceed differently and actively seek out those things with which I disagree. Why? Because if I disagree with the science then one of us is wrong and as a scientist my search is for consistency and ultimately what can be verified by experiment. Part of what scientists do is read papers that draw conclusions with which they disagree. In my own case I then try to understand the flaw in the paper's argument, and failing this, I try to identify the mistakes in my own reasoning that led me down the garden path. Of course such detailed analysis often shows that we were both partly right and partly wrong and the clarifications lead to deeper understanding.

The reader is probably not interested in how scientists generally carry out their research activities, so I do not discuss those activities here. The presentation is concerned with explaining the scientific foundation of three concepts: uncertainty, inequality and unfairness. I will side step the temptation of defining these concepts here and assert that all three emerge from the complexity of phenomenon in the physical, social and life sciences. Since I am neither a physician nor a social scientist the reader is certainly justified in asking how I have come to some of my conclusions, particularly those that differ from the opinion of a large segment of society. This is the reason I begin with a discussion of how scientists in general think when they are attempting to understand complicated problems, or more specifically how I think about complexity. Since this is not a scientific publication I freely express my opinions along side what I can prove, with the self-imposed constraint that I explain how I formed the opinion expressed.

I am a physicist so I tend to think about solving problems in a particular way and this gives me the advantage of method. I apply the same method I use in my work to such questions as the existence or non-existence of certainty, equality and fairness in society. I start Chapter Two from the premise that equality and fairness are a consequence of simplicity and the entire chapter is used to explore the consequences of that hypothesis and explore the evidence in support of it.

A simple process can be predictable or not. For example, flipping a coin is simple but not predictable, whereas tossing a horse shoe is both simple and to a large extent predictable. There are two ways a process can be unpredictable. In the case of a coin toss the process is random and therefore is by definition unpredictable.

The second way to lose predictability is by increasing the complexity of a process, which we take up in Chapter Three. Therefore a phenomenon need not be complex to be unpredictable; it only needs to be random. It is worth mentioning here that a simple random processes is one described by Normal statistics and is described in great detail without mathematics in Chapter Two. In that chapter the properties of Normal statistics are shown to be the basis of many of our modern ideas including certainty, equality and fairness.

On the other hand, complexity is measured by the deviation of the statistics from the familiar bell shape. The bizarre properties of these statistics are discussed in Chapter Three. These new statistics are used to describe phenomena dominated by events out in the extremes; stock market crashes, earthquakes, floods and other extrema that determine the drama in our lives. These statistics are descriptively called heavy-tailed to capture their emphasis on extreme values. Random fluctuations described by these inverse power-law statistics represent processes that are intermittent in time, like the splashing drops from a leaky faucet, or processes that cluster in space, such as the formation of spontaneous traffic jams. The entire content of this book is to provide a rationale for those events that can have a significant impact on daily life and how we think about them, while being guided by the difference between bell-shaped and heavy-tailed statistics.

My oldest son has always thought the world ought to be fair. When he was eight or so he noticed during a class that his teacher kept glancing back and forth between arithmetic problems she was solving on the blackboard and a sheet of paper on her desk. Straining forward from his front row seat he was able to determine that the paper had all the problems worked out in advance. His response was to stand up in front of the class and reprimand her by saying: "Miss __ you are cheating". The resulting student-teacher interchange was the topic of an impromptu parent-teacher conference later that day.

This incident occurs to me now because it was such a clear and personal experience of how preconceptions determine the ways we interact with one another. My son's concern for fairness overruled his sense of good manners and judgement about classroom behavior. But then he was only eight and to be fair he was also right. Of course thinking about an incident of this kind and deciding to put it in a book are very different things. My reason for writing about it is to emphasize that his view of the world and fairness, which he still holds some thirty-five years later, are very different from my own. The question addressed in this book is not whether the world is fair, the evidence clearly shows that it is not, but whether it ought to be fair. The answer to the latter question is not so obvious.

There are many ways to address questions of what ought to be true about the world. There is the philosophical, where one can draw from the great thinkers of the past and construct impressive arguments complete with footnotes; there is the strictly theological, where one accepts a few uncontested truths to start and from them draw a series of logical conclusions; and finally there is the historical in which one can trace what has occurred in the past and argue that this is what will occur in the future. These approaches and many others have a common failing in that explaining complex phenomena, such as how wars begin and end, how economies are destabilized, or why couples divorce, involve ignoring and/or suppressing the very features that make them complex or assuming they cannot be known. This is the most common method of simplification, leave out those annoying details that detract from the main message of the person doing the explaining.

Complexity is not merely complicated simplicity and simplifying complexity is not the same as ignoring the properties that make a phenomenon complex. What I hope to make reasonable is that many complex phenomena are not understood because they are not properly represented. In the proper representation what happens in even very complex events seem natural and often even predictable, but such representations are not obvious and are frequently counter-intuitive. Thus, finding the proper representation is an adventure in itself.

On the other hand, there is the more disruptive situation in which the apparently simple is shown to be complex. This is the situation where the zigs and zags of reality have been smoothed over using a simple map that explains things in the way we choose to see them. Very often it is familiarity that gives the illusion of simplicity and leads to misunderstandings. For example, in the United States the view that the slave economy of the antebellum south was unprofitable, stagnant, inefficient, and moribund was wide spread according to the 1993 Noble Prize winner in economics Robert William Fogel [1]. Fogel corrects the historical missimpressions regarding slavery as follows:

> "...the new economic historians have demonstrated that slavery was quite profitable. To put it in contemporary terms, as an investment opportunity, slavery was the growth stock of the 1850's. Thus when slaveowners invested in slaves, it was not because they were doddering idiots wedded to an economically moribund institution. Nor was it because they were noble men who were sacrificing their personal economic interests to save the country from the threat of barbarism. Perhaps slaveowners were nobly motivated. If so they were well rewarded for their nobility - with average rates of return in the neighborhood of 10 or 20 per cent per annum. New measurements also indicate that the slave economy was growing between 1840 and 1860. Far from being the laggard region, the rate of growth of per capita income in the South exceeded the national average. But perhaps the most startling of the new findings is the discovery that Southern agriculture was nearly 40 per cent more efficient in the utilization of its productive resources than was Northern agriculture."

Consequently it is not only accurate simplification of the complex that we pursue, but uncovering the complexity hidden beneath previously established, but misguided simple cognitive maps also concerns us. The latter is important because it is only by revealing the underlying complexity that we can hope to understand phenomena well enough to make them simple again and through understanding control them.

Much of our technological society appears to be simple because what makes it work is hidden under multiple layers of technology. A modern city cannot function without transportation networks for logistic support of its markets, sewers and water delivery networks for hygiene, the power grid, communication networks, and on and on. Each of these networks is in itself a complicated interconnected system of dynamic elements that is organized and/or designed to carry out specific functions. When the city is operating as intended these various networks are part of the background and are intended to function invisibly. On the other hand when one or more of these networks does not function correctly such as having an increasing number of homeless, skyrocketing health care costs, incredible gasoline prices and so on, the background becomes the foreground and we question why the city, county, state and federal governments cannot solve the social problem. This book is not about why we cannot solve these problems. It is about why we see certain things as problems in the first place.

Why are so many people poor? The existence of a poverty class is perhaps understandable in Third World countries, but why in the richest country in the world are there so many poor? Can't we do something about it? The simple and direct answer to this social problem is that through the equitable redistribution of a nation's wealth we can end poverty. But will giving everyone an equal share of a nation's wealth end poverty? Or is this a pleasant myth created, spread and accepted because of our limited understanding of how our complex institutions work?

The idea of fairness is recent in human history and has slowly evolved in the classical writings of the last few hundred years. It took on much of its modern form in the nineteenth century when scientists turned their attention from the understanding of physical to that of social phenomena. The mathematics that made this transition possible was the introduction and development of statistics, which enabled scientists to identify the common aspects of apparently random data sets from a variety of phenomena. It should also be recalled that the mathematics of statistics was developed to predict the most favorable outcomes of games of chance in which the notion of a fair bet was central to the understanding of a wager. Recognizing the uncertainty in the outcome of human interactions the nineteenth century social scientists adopted statistics and probability as the proper calculus for describing social phenomena. This choice of mathematical representation introduced a number of foundational concepts into our understanding of society; these include equality and fairness.

1.1. ONE SCIENTIST'S VIEW

My approach to understanding the complex issues of today's world is that of a scientist. When I was younger and attended parties I responded to such questions as: "What do you do?", with the ill-conceived response: "I am a theoretical nuclear physicist". Invariably the questioner's eyes would glaze over as they furtively looked for a way to be anywhere else in the room. I am now older and wiser and now my response is: "I am a scientist." For some reason the word scientist less often evokes the 'fight or flight' response that theoretical nuclear physicist invariably did. Occasionally people even stay and talk with me. However my interests are more in how people form the opinions they want to share than in what those opinions actually are. That shift in perspective from what they think to why they think it apparently offends some people who consider it aggressive and therefore even those that stay to chat don't stay long.

It is natural for a scientist to want to understand the basis from which people draw conclusions. It is one of the few things most scientists have in common, regardless of whether they are trained in biology, chemistry, physics or sociology. In many other respects scientists are as individualistic as any other group. Consequently, my view of the world as a scientist also has a thick layer of ego superimposed; like most other people I understand even the most formal aspects of the world in a personal way. To clearly see why I approach the understanding of complex issues and their simplification the way I do can be presented in at least two different ways. I could present the reasoning supporting my position, and some would say that should suffice. But in my experience such a dry discourse is only sufficient for another scientist and one who is already sympathetic to that point of view. Most people want to experience an underlying story or a passion to which they can relate. So I choose the second approach and interleave how I formed my opinions regarding equality and fairness in society with the science supporting those opinions. So let me start with graduate school.

The Vietnam War circumscribed my days as a physics graduate student. In classes, coffee shops, or walking on the university quad one bumped up against this reality and the position one took on the war, for or against, defined your persona and how you were treated by your peers. I observed first-hand the vicious arguments between the pro-war and anti-war factions; rhetoric that inflamed audiences, but did little to persuade; Friday night rallies and Saturday afternoon marches; long-time friends that stopped talking to one another, were all part of the social framework of my studies. Out of this activity one thing became increasingly apparent over time; the arguments had less to do with the war and more to do with how government authority fit into an individual student's world view. Was the military draft legal? Who should receive exemptions from the draft? Was the lottery fair? Should those that avoided the draft be given immunity? These and many similar questions were answered in idiosyncratic ways, often forcing the questioner to re-examine other related issues in order to arrive at an acceptable level of intellectual comfort. While I think it is possible for human beings to simultaneous believe in the truth of diametrically opposed propositions, which in friends is seen as broad-mindedness and in enemies as hypocrisy, intellectual consistency is seen by most as being more desirable.

In the years since I first thought about these things it has become increasingly clear to me that the psychologists are correct in their observation that people establish their view of the world at a relatively young age, but at exactly what age is not clear. Some believe that the basic outline of a mental map is achieved in the first few grades in school, or even in pre-school pointing to parental influence. But I see that young adults in high school and college are still malleable and responsive to outside influences albeit to varying degrees, depending on how sensitive they are to peer pressure. The experiences as a child and young adult mould and shape the way we represent the world to ourselves, some experiences being foundational and others merely cosmetic. However even graduate students can be strongly influenced by teachers with radically different world views and empathy coupled to an ability to communicate. My own transformation was due to a man who touched my nascent vision of how the world might be understood if only I knew a great deal more. I learned that understanding comes in stages and can be developed using only the tools available. No reason to wait.

Elliott Water Montroll (1916-1983), the then Einstein Professor of Physics at the University of Rochester, was one of those rare individuals who appreciated the importance of nonlinearity in complex systems. Elliott had academic training in chemistry, mathematics and physics, but had research interests that transcended traditional disciplinary boundaries. He was a giant in the development of mathematical models to describe the statistical behavior of complex dynamic systems, but his friends found his insights into the models that he revealed in casual conversation even more fascinating. He once compared what he did with the role of a country doctor, who had a black bag filled with mathematical methods and physical models, that could be pulled out and applied to whatever scientific maladies he encountered. He had a way of viewing the world that enabled students and colleagues alike to appreciate the quantifiable within the most qualitative aspects of social sciences, particularly those that involved technological evolution.

Let me present an example of his thinking paraphrased and updated from his last article [2] published after his death, which involves using the increase in the distance a typical worker can travel on a day's wages as a measure of affluence. Some 1700 years ago, according to Diocletian's wage and price control ordinance, an unskilled workman could travel about eleven miles on a day's wage, while a carpenter or stonemason could do twenty-two miles. That is the way things stood for the next 1500 years. In England toward the end of the eighteenth century the stagecoach fare from Manchester to London, a distance of 195 miles, was four miles to the shilling. A laborer's daily wage was slightly more than a four-mile deluxe ride. He could do about as well as the Roman by renting a horse. A coal miner did not do so well, but a foreman could travel commercially about seven miles on a day's wages. While the coaches were the best vehicles available, they did jostle the passenger mercilessly. In New England the tavern density was one per linear mile to ease the traveler of his pain.

Elliott also points out that in 1795 the stagecoach fare from New York to Georgetown, the present day location of the nation's capital, was divided into three stages: New York to Philadelphia at a cost of $6, Philadelphia to Baltimore for the another $6, and Baltimore to Georgetown for $4. This total of $16 is to be compared with the 2011 bus fare of $21 on weekdays and $25 on weekends for a smoother four and one half hour ride that is temperature controlled. Typical colonial stagecoach fares in non-mountainous, well-settled regions ranged from five to seven cents per mile, rising to ten cents per mile in wilder parts of the country. The government travel allowance was fourteen cents per mile in 1815 and today it is forty-seven cents. The numbers reveal that if a poor man needed to take a long trip he probably walked. Those slightly better off might own a horse, but only the rich could afford commercial vehicles. A European without funds who wished to try his luck in the colonies had to indenture himself for seven years (about one-third of his remaining life expectancy at age twenty) to pay for his boat ride.

The first technological breakthrough to broaden the modern worker's travel horizon on a day's wage beyond that of the Romans was the construction of canals. When the Duke of Bridgewater's canal was completed in 1765, the price of 5 pounds sterling per ton from Liverpool to Birmingham was reduced by 80%.

Canals were a basic component in the birth of the industrial revolution [2]. The early steam engine provided the motor power to make possible the cheap mass production of simple manufactured objects. However, mass production had little social value without mass markets. Before canals, transport costs were frequently greater than production costs, so that a change in production cost had only a secondary effect on the price at a distant market. The canals at low cost carried coal and raw materials for fabrication to the industrial centers and then cheaply delivered finished products to distant markets. By the turn of the nineteenth century hardly any significant English town was more than ten miles from a canal.

A more dramatic change in price, trip time, and comfort came with the railroads, which replaced the canals. The railroad fare in the period 1890 to 1915 was about two cents per mile. Hence, on the $5 a day that Henry Ford paid his workers, they could travel 250 miles. A UAW auto worker in 2007 made about $320 per day including wages, bonuses, overtime and paid time off or approximately $75,000 per year. Today, 2011 regular airfare is about 23 cents per mile in the United States so that a trip of approximately 1400 miles is possible by air on a day's wage. With careful planning, taking advantage of advance purchasing on the Web a flight from New York to Los Angeles on Southwest Airlines costs 6.8 cents per mile, whereas Jet Blue costs 4.8 cents per mile. A skilled industrial worker could then fly across the United States and back again on a day's wage. His poorer brother who got along on minimum wage (average of $7.50 an hour) could still travel 1250 miles by Jet Blue, a factor of 115 times better than the old Roman and five times better than Henry Ford's employees. These numbers are somewhat exaggerated since we did not take taxes into account in the calculations. Of course estimating the average cost of a gallon of gasoline to be on the order of $4 a typical worker could drive his car with a wife and two children across the country and back again for a day's wage.

The key to the above discussion was deciding on the measure of affluence, how far a worker can travel on a day's wage, and then exploiting the existing data to determine what we can learn using this measure. Different measures might lead to different conclusions so we have to agree at the beginning that the measure we decide upon truly measures the quality of interest. Often the most probative measure is not apparent and we have to develop theories and/or metaphors to suggest how to connect available data to the qualities we wish to understand.

I use a variety of metaphors to describe how our past experiences determine our interpretation of present events. Don Quixote saw the world in terms of virtuous maidens worthy of protection and grand challenges to be met and overcome. By contrast Ebenezer Scrooge was a dark soul whose humanity had been suppressed to the point where he saw human interactions only in terms of money, not for the luxury it could bring, but to keep score on what was owed him. Whether we are sympathetic to these fictional characters is determined in large part by our acceptance or rejection of their formative experiences and the cognitive map they constructed for themselves as revealed by the author through narrative. This is my favorite metaphor, cognitive cartography, and is one I return to again and again.

The term cognitive map was originally used in a technical way as a method to construct and accumulate spatial knowledge that facilitated learning and enhanced recall. I use it in a much broader metaphorical way to represent an individual perspective of the world; not just on the physical placement of things, but the judgement, feelings and associations with those things as well. I want to avoid the tedium of definitions, so like most of what we learn in life the meaning of cognitive map and other such jargon is clarified by how we subsequently use them.

1.1.1. Changing Perspective

A cognitive map is not static but changes over time, with perhaps even the most deeply held beliefs being challenged and modified by intense new experiences. Even the most senior citizens can have transforming experiences; one of the more optimistic tales of such transformation was given by Charles Dickens in his account of Scrooge in *A Christmas Carol*. For many people the attraction of great literature is that it allows them to re-examine their core beliefs at a safe distance regardless of their age. Another is that fiction allows the interesting depiction of deep truths that outside the context of a story seem either contrived or self-evident.

Science has many similarities to literature. It is a discipline that enables us to catalogue and organize the basic facts of our lives as we store them as data. But like a closet that is nearly full, each newly stored item (piece of data) requires reorganization in order for it to find a place for long-term storage. Some items generate reorganization, but in the end they cannot be made to fit and are discarded. Thus, the reason for some reorganizing is not contained in the map. But whether the cause is there or not, reorganization is done at a higher level of abstraction than simple data collection and is accomplished through identifying patterns within the data, or by imposing patterns on the data that we believe ought to be there. We can remember an arbitrary number of words in a story or poem, but nonsense words strung together slip out of memory almost immediately. It is the coherent pattern made through word associations, the story, that make certain strings of words memorable.

Patterns seen in different contexts can be associated with one another; the young buck's challenge for leadership of the herd becomes the Oedipus Complex in human psychology and the alpha male leader of the pack becomes the great man theory of history. Or perhaps the other way around. Those that make such associations for the first time are considered exceptional and those that do it systematically either become artists or scientists. In physics the beauty of the multicolored arch of a rainbow can be isolated in a laboratory using a prism. The spread of colors on the laboratory wall may lack the majesty of a rainbow's backdrop, but when seen for the first time it can be just as breathtaking. In science such patterns can be thought of as information and it is this information that interrelates and organizes various pieces of data within the cognitive map. A great many people are satisfied with this form of the map; for them the pattern of violence experienced as a child 'explains' the pattern of predatory behavior of an adult. For others this pattern association does not 'explain' the behavior. In particular the physical scientist does not view the awareness of the patterns alone, no matter how detailed, as explaining the phenomenon.

Sir Isaac Newton first analyzed sunlight by shinning it through a prism in his laboratory and discovered the same phenomenon he and countless others observe after a summer shower. He found that the white light was not a single color, but consists of all the colors of the rainbow stacked together and his prism enabled him to pull them apart. His theory predicts how the different wavelengths in white light are bent by differing amounts upon entering and exiting a prism and provides knowledge that does not mar the beauty of the rainbow, but enhances it because the mystery behind the fan of color is revealed. This theory was violently rejected by the poet and scientist Goethe in his *Theory of Colour*. The poet in Goethe could not make compatible the aesthetic principles associated with the effects of color and the experiments of Newton and so he resolved to carry out the experiments himself. He did not believe that light could be understood outside the garden, by which he meant that phenomena had to be understood in the context in which they were found and not in the laboratory. He did in fact reproduce the experimental results of Newton, but he did not reach the same conclusions, but critiqued Newton's theory with phrases such as: "incredibly impudent"; "mere twaddle"; "ludicrous explanation"; "admirable for school-children in a go-cart"; "but I see nothing will do but lying, and plenty of it". These observations and more may be found in *Popular Scientific Lectures* by Hermann von Helmholtz in a lecture on Goethe delivered in 1853, who perhaps best summarizes Goethe's view as:

> "Just as a genuine work of art cannot bear retouching by a strange hand, so he would have us believe Nature resists the interference of the experimenter who tortures her and disturbs her; and, in revenge, misleads the impertinent kill-joy by a distorted image of herself."

A scientist wants to extract knowledge from information and such knowledge only comes about through the formation and application of theory. Science therefore consists of three parts: the acquisition of data; the identification of patterns (information) within the data; and the formulation of logical structures to interpret information (the patterns) and obtain knowledge (theory). The cognitive map of the scientist reflects this three-fold partitioning. However, history teaches that science constitutes only one form of knowing; the arts and humanities constitute a different reality that science with its instruments and predictions may not be able to probe. However rather than entering into this hoary debate I will attempt to finesse it by simply explaining various parts of my own map and letting the reader decide whether there is really more than one way of knowing.

1.1.2. Grand Visions

My particular prejudice is that all knowledge involves theory, and it is only through experiment and the subsequent explanation of experimental results by theory that we are able to make sense of the world in which we live. From this perspective I see western society as being dominated by a particular cognitive cartography that influences how we think about and understand our world. This particular notion did not originate with me, but with some of the great social thinkers of the nineteenth century.

Adam Smith (1723-1790) was brilliant, absent minded, talked to himself, and is credited with being the father of modern economics and capitalism. He was a social philosopher who in his 1776 book *The Wealth of Nations* [3] invoked an invisible hand that served society through individuals seeking their own self-interest. This work represented a shift in economic thinking no less significant that Darwin's *On the Origin of Species* in biology and Newton's *Principia Mathematica* in physics. The economic theory of Smith was strongly filtered through the disinfectant of Christian theology and the 'invisible hand' in economics was no less plausible than 'the hand of God' in the workings of an individual's life.

Another social thinker, Karl Marx (1818-1883), was born into a wealthy Prussian family, and died a stateless person in England. He is perhaps best known for his

1848 work *The Communist Manifesto* [4], which he coauthored under the influence of his friend F. Engels. Throughout his writings Marx professed that the mode of production was the social force that drives human history. The value of individuals to society was the result of their contribution through the production of wealth by means of manual labor. Capital is the means by which labor is controlled and it is only through ownership of capital by labor that the wealth of a nation can be fairly distributed. In this theoretical system all people are truly equal.

Thomas Carlyle (1795-1881) was an irascible social commentator. In his master work *On Heroes, Hero Worship, and the Heroes of History* [5] he formulated the great man theory of history. In this theory it is by the great man imposing his will on the nearly random events of history that direction and form emerge out of chaos. Without great men, history would cease to have purpose, society would have no more historical significance than the social gatherings of monkeys and baboons.

Each of these three contributions provides a different lens to observe and interpret the behavior of human beings and the societies they form. Each originator presented detailed descriptions of the patterns they saw in the human record and provided a great deal of discussion regarding the explanatory value of these patterns. However in no case was there an underlying scientific theory, because the criterion for what constitutes a scientific theory is the ability to make predictions that can be tested by experiment. The economic predictions, either on the very small or on the very large scale, fall short in this regard. Only within the last quarter century or so with the convergence of economics and psychology has neuroeconomics been able to predict the economic behavior of small cohorts of individuals. In each of the historical examples the phenomena being 'explained' were too complex for the 'observed' patterns to be predictable. The patterns seen by the giants of social science are the result of careful organization of some data and the total disregard of other data.

The mathematician Norbert Wiener (1894-1964) in his book tracing the historical and sociological significance of ideas in the context of inventions [6] put it this way:

"Kipling has emphasized ....English patriotism centers about the king, and American patriotism about the flag. "There is too much Romeo and too little balcony about our [the English] National anthem", he says, "With the American article it is all balcony" ...The part that the king plays is that of Romeo, and the part that the flag plays is essentially that of a drapery for the balcony.""

The economic theory of history is all balcony and no Romeo. On the other hand the great man theory of history is all Romeo and no balcony.

These theories of history could not be more different and the implications drawn from them have determined the world's trajectory over the past century, or so it would seem. Rather than seeking some overarching social force I propose to determine what can be learned from the general thesis that society is complex and measurable. The path I follow in this book is the one less travelled and it diverges from the traditional in that I look for the complex within familiar natural and social phenomena. I examine the patterns in the making of wars, in organizing or joining a strike, in the beating of the human heart, in breathing, in walking, in laughing at a joke, or in any of an increasingly large number of other phenomena that seem simple when viewed from a distance. However, even the simplest of things, when examined close up reveals an intricacy that is often avoided in discussion because of a lack of words with which to explain their striking complexity. For example, there can be sudden changes in a process that are not anticipated by past behavior; the unexpected shiver down my back when I step out of the ocean into the night air, or the anticipated, but unpredictable, time of my next hiccup, are examples of everyday phenomena that do not lend themselves to simple predictive descriptions. These things do not seem complicated, but each of them is complex in the sense of being unpredictable, but not totally random.

One way to understand the world, at least at first, is through simple generalizations. A child believes the world is friendly and jumps into a stranger's arms; another feels less certain and approaches people with more caution; still others find the world to be hostile and reject the overtures of nearly everyone. The mental maps are very different for these three categories of children and their subsequent development will in all likelihood be different as well. As they grow each identifies familiar patterns that subsequently reinforce what they already believe, unless an intense experience imposes a new pattern to be used for comparison. It is evident that the world view constructed in this way has certain arbitrary features that may, or may not conform to what the world is really like. So what do scientists do that is unique and that enables them to form a more faithful mapping of the world?

A scientist is a person that believes that the confusion in the world can be clarified through reason; phenomena can be understood through a judicious balance of observation, experiment and theory, and surprises can be predicted and therefore avoided. In addition to what other human beings do to reduce the confusion in their lives the scientist tries to understand the world through the use of simple models. Models are abstract constructs, often mathematical, in which the symbols refer to measurable properties of interest. Mathematical analysis replaces verbal reasoning, as might be given by a lawyer or philosopher, and enables the modeler to draw inferences about how the world would behave if only it satisfied the assumptions made in constructing the model. However, I do not want to go on a professorial rant here, concerning the care and feeding of scientists and their models, but I do want to indicate what models contribute to how a scientist answers questions for himself and for others starting from when they were very young.

A child who derives more pleasure from uncovering the deception in a magic trick than they do in the illusion itself is a candidate for studying science. My wife took our two sons to see the musical *Annie.* Afterward she told me that a large number of scene changes were done with mechanical devices moving the stages around. She was disappointed with the show, but our sons were not. She thought the singing and acting were not at a professional level. Whereas our sons mostly ignored the actors singing on stage and had a wonderful time figuring out how the stage mechanisms worked. One son is now a physicist and the other has done spectacularly well in designing and developing computer games. Both sons have a strong analytic orientation, meaning that they enjoy figuring stuff out for themselves.

Science begins by acknowledging ignorance and suspecting the word of experts;

accepting that one is on the frontier between what is understood and what is not. Not understanding stimulates a curiosity about the details and depth of what is not known. And finally the curiosity is manifest in the formulation of questions whose answers can fill in sections of the cognitive map and reduce the area of the unknown. The exploration involved in answering such questions is the self-assigned task of the scientist. But the questions are not always formulated by the scientist. Very often a profound question is supplied by a lay person such as a politician: What can governments do to eliminate poverty, hunger and the fundamental imbalance in the distribution of wealth?

A related question has to do with the growth of Earth's population and what if anything we can do about it. One of the things a politician might want to understand and influence is how populations grow over time under changing environmental and social conditions. Consequently they might turn to a scientist and ask: Is the planet in danger of becoming so overcrowded that the human species will destroy itself?

But when you ask a scientist a sober question you need to be prepared for a detailed, sometimes annoyingly qualified and perhaps overly long answer. As a scientist I know that in order to begin thinking seriously about answering a question I have to determine what science already knows in terms of the available data. Answering this question on overpopulation provides a context in which to show how using models enables a person to think quantitatively. How to address overpopulation was also a popular question when I was a graduate student in the late 1960s and its understanding was one of the topics pursued by Elliott Montroll.

1.2. POPULATION MODELS AND QUANTITATIVE REASONING

In the background of most scientific thinking is mathematics. The formalism of mathematics is the medium scientists use for self-expression and its selection is no less important than the medium selected by an artist. Some use oil paints (the differential calculus), others use water colors (number theory), while still others use clay and marble (geometry), but the purpose is always the same and that is to gain deeper understanding of the phenomenon being studied. One artist uses a broad palette drawing from many kinds of mathematics, while another restricts herself to the primary colors and explores the depth of the medium along with the phenomenon of interest. Words are very often ambiguous and masterful writers can exploit that ambiguity to stimulate the imagination in telling powerful stories. The more a story can evoke or awaken the experience of the reader, the more successful the work. It is this very reason that scientists intentionally narrow their prose, choosing to express the most imaginative part of their activity through the clarity of the mathematics.

It is the common aspects of the panic that ruled the floor of the stock exchange in the early part of the last century and the joy of a ninth inning winning run by the home team, that when stripped of emotion enables the scientist to more fully understand the psychology of consensus. The mathematics describing the phenomenon of agreement is devoid of emotion, but the interpretation of the mathematics is an entirely different matter. The theory necessary to extract knowledge from information patterns is constructed using a mathematical infrastructure overlaid with connective interpretation. However a theory is an integrated whole and the mathematics cannot be usefully separated from the interpretation to facilitate presentation. What can be done is the mathematics can be transformed to a more familiar representation for understanding and then it can be revealed how the interpretation is teased from the data. As a shorthand for this activity I use the terms model and modeling and in its discussion I avoid the use of equations. For me and probably for most physical scientists this is a lot like talking with my hands tied behind my back; so to provide a bit of flexibility I replace the mathematical equations with their solutions in the form of pictures and graphs. Hopefully the curves, along with the discussion of their ups and downs, will provide sufficient insight into the models with the various bendings being associated with the underlying mechanisms that induce the changes.

What a scientist strives to construct is a calculus of everyday phenomena or more precisely a way to think quantitatively about everyday phenomena. This book strives to be a primer on how to use the scientific method, or at least one version of that method that can be used by the non-specialist, to think more clearly about those aspects of the world over which they have some control. In order to achieve this goal it is not necessary to identify what we do not know, that is simple and consists of the majority of things we encounter. What is necessary and difficult is to identify what we know to be true that is wrong.

Elliott Montroll and his graduate student Wade Badger, who I roomed with for a while in graduate school, in their 1974 book *Introduction to the Quantitative Aspects of Social Phenomena* [7] made some of the first connections between mathematical modeling and such things as the growth of populations, the arms race, various kinds of statistics, speculation and the stock market, pollution and the growth of cities. It was not that they were the first to address modeling some of these phenomena, they weren't; but they did make an effort to view them all from a single coherent perspective, separating the simple and readily grasped from the complex and less easily understood. In the latter category could be placed such contemporary questions as: What does a Dutch tulip bulb of the early seventeenth century have in common with the dot.com business of the late twentieth century and the new science of econophysics? To answer this question requires a reorientation of how we think about ourselves and our world. To begin this change of how we arrange things in our head we start where these two early pioneers into understanding social complexity. Montroll and Badger, started and that is with the growth of populations. We do this because such growth is interesting in itself and modeling it provides insight into a kind of complexity that can be developed in stages and transferred to the understanding of other complex phenomena.

1.2.1. Exponential Growth

The population of Europe grew at a slow constant rate for about the last two millennia and the rate increased through the nineteenth century. By the end of the eighteenth century a number of people had observed that the European population doubled at regular intervals, a phenomenon characteristic of exponential growth. The cleric Thomas Robert Malthus (1766-1834) is often credited with this observation, but many others, including Thomas Jefferson had made similar observations. It is noteworthy that the most cited work on population growth was by a cleric writing a discourse on moral philosophy. However Malthus did draw some dark conclusions that were later vindicated [8]:

"If the United States of America continue increasing, which they certainly will do, though not with the same rapidity as formerly, the Indians will be driven further and further back into the country, till the whole race is ultimately exterminated and the territory is incapable of further extension."

Malthus was exploring the consequences of the fact that an exponentially growing population always overtakes and exceeds a linearly growing food supply. The result is overcrowding and misery. He stated [8]:

"Population, when unchecked, increases in a geometric ratio. Subsistence increases only in an arithmetical ratio. A slight acquaintance with numbers will show the immensity of the first power in comparison of the second."

The fact that there was no evidence that the food supply was growing linearly was overlooked; this was merely a convenient assumption for Malthus. An assumption that led to the exponential population growth depicted in Fig. (**1.1**) eventually overcoming the linear growth in the food supply allowing him to make his point about the growth of human misery. Putting aside the moral considerations for the moment let us look at the apparatus moving the stage around and ask: What is the real difference between linear and exponential growth?

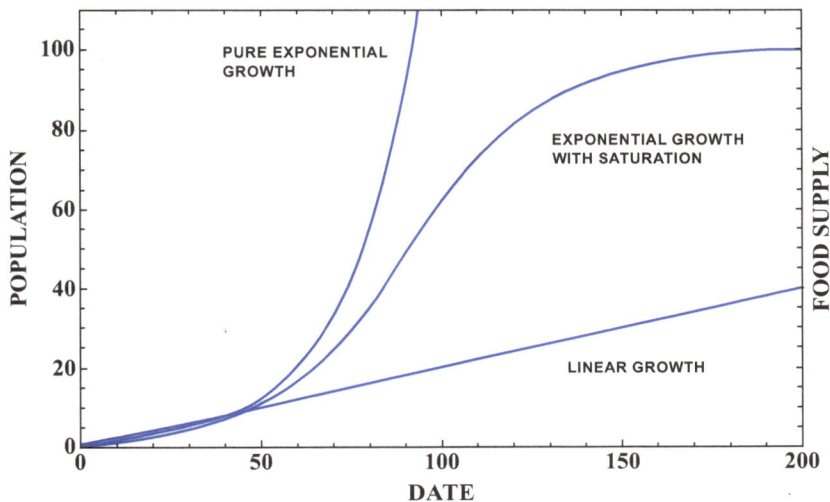

Fig. (1.1). The general form of the diverging exponential population growth of Malthus along with the linear growth in the food supply compared with the saturated population growth of Verhulst. Note that the units for the size of the population, food supply and date are arbitrary, and can be adjusted to accommodate specific data sets.

Linear growth means that the amount of stuff (system size) increases by the same amount in each time interval. Suppose in the first year the size of the system is 1, then in the second year the size of the system might be 2, in the third year it would be 3 and so on. Therefore after an arbitrary number of years, say n, the total size of the system is n. Thus, if the size of the harvest is a million bushels of corn in the first year, it will be ten times that size a decade later. This steady increase in the supply of food is maintained by clearing more land for planting each year, increasing the efficiency of the land in use through fertilizers and irrigation, developing technology for more efficient ploughing, planting and harvesting and so on. So the assumption of linear growth in this context is not trivial and to be maintained requires a great effort on the part of society and the farmer.

The next question is whether the exponential population growth predicted by Malthus is really as dire as he believed? A reality check might be in order here. Suppose I were to tell you that I would work for you on the following pay schedule. The first day pay me one penny, the second day two pennies, and the third day four, always doubling the pay of the previous day. Is this a good deal for you or for me? You know that eventually I will make a lot of money, but how long will I have to work? On the last day of the first month (30 days or 30 doublings of the penny) my pay will be over a million dollars. What kind of idiot would continue to pay me according to our prescription? I know that you would not. The last day of the second month (60 days or 60 doublings of the penny) my pay will be larger than the National Debt of the United States. So you can see that exponential growth plunders resources to maintain itself and in the real world it cannot be sustained. Of course in the real world unqualified linear growth is also a fiction.

Exponential growth in one variable can be compared to linear growth in a different variable. In linear growth the increase from one year to the next is the same generating the sequence of sizes, 1, 2, 3, ... Consequently there is a unit increase in size in each year from the preceding year. In the exponential case it is not the increase that remains the same from one year to the next but the percent increase that remains the same. For example in a slow exponentially growing process, the sequence of size increases might be 1%, 1%, 1%, ...Compare this sequence with that of the linear growth: 100%, 50%, 33 1/3%,..., which yields a decreasing proportion each year. The linear growth starts out much faster than the exponential, but it is eventually overcome, because a decreasing series is always eventually overcome by an unbounded series of constant factors. In the pennies a day example, with a factor of two increase from one day to the next, the percent increase is the same large constant from day to day: 100%, 100%, 100%, ... This doubling overtakes linear growth in one day.

Kenneth E. Boulding [9] formulated the Malthus view of the misery resulting from unchecked population growth as the *Dismal Theorem*. He in fact formulated three versions of the theorem:

> " *The Dismal Theorem:* If the only ultimate check on the growth of population is misery, then the population will grow until it is miserable enough to stop its growth."

> "*The Utterly Dismal Theorem:* Any technical improvement can only relieve misery for a while, for so long as misery is the only check on population, the [technical] improvement will enable population to grow, and will soon enable more people to live in misery than before. The final result of [technical] improvements, therefore, is to increase the equilibrium population which is to increase the total sum of human misery."

> "*The moderately cheerful form of the Dismal Theorem:* If something else, other than misery and starvation, can be found which will keep a prosperous population in check, the population does not have to grow until it is miserable and starves, and it can be stably prosperous."

Boulding was not too enthusiastic about being able to prove the third theorem.

If misery in the form of poverty is one side of the social coin then happiness in the form of affluence, must be the other. Montroll's measure of affluence introduced earlier was the distance a typical worker can travel on a day's wage; another measure he used was the amount of goods purchasable by a day's wage. He [2] made this measure systematic using the ratio of the index of the industrial daily wage to the wholesale farm price index for the period 1860-1975 (with the ratio for 1865 being arbitrarily set equal to one). The use of price indices may seem unnecessarily academic, but they do avoid the impression that the observations made depend on the value of currency at some arbitrary point in history, which they do not. The farm price data corresponds to the same mix of produce for each year of data recorded. The increase in the index by an order of magnitude depicted in Fig. (**1.2**) is the result of improved agricultural efficiency and of the increased productivity of a typical non-agricultural worker so that he can command a higher wage. Consequently Montroll shows the effect of technological development on affluence in the United States in the past hundred years or so and how this has mitigated Malthus'prediction of misery.

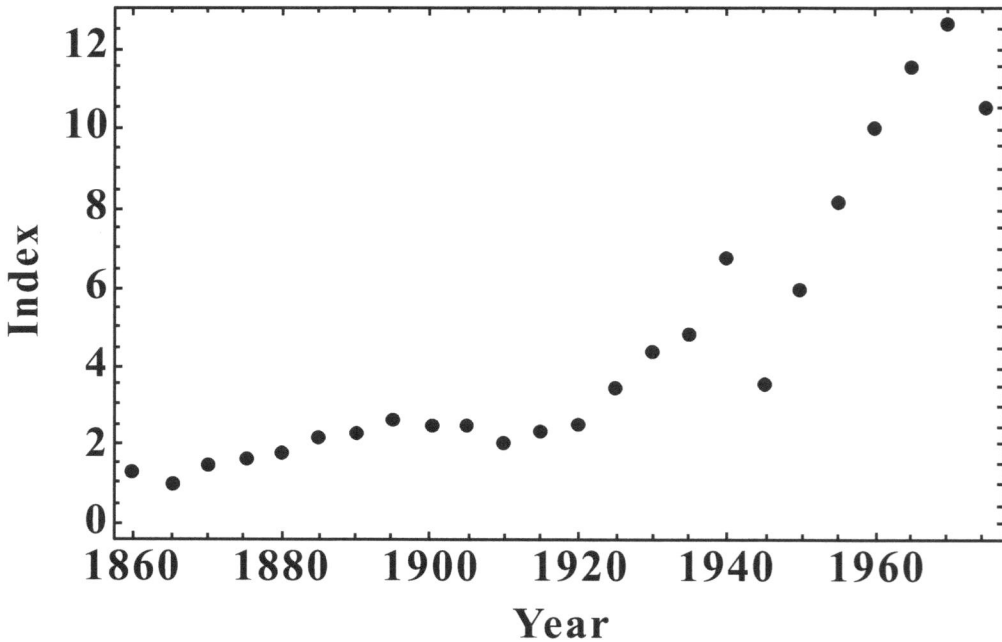

Fig. (1.2). Index of the ratio of industrial daily wage to wholesale farm price index 1860-1975 (Ratio 1865 = 1.00; data from [2]).

We are also in a position to check the assumption made by Malthus that the growth in the availability of food is linear. Montroll [2] recorded improved agricultural yields in wheat, corn and cotton shown in Fig. (**1.3**) for the United States since 1800. The measure recorded in the graph is the ratio of how long it takes a man to work an acre of ground (man-hours/acre) and the crop yield that is returned for his effort (bushels/acre). This amazing reduction in the time required to produce a bushel of yield is a consequence of advances in agricultural technology. It is the result of the proper choice and use of fertilizer, research in plant genetics, mechanization, improved soil management, irrigation, and other factors. Modern agricultural evolution has been a continuing replacement of one technique or plant strain by another for a very long time. The result is a spectacular linear decrease in the expenditure of man-hours per bushel over this 150 year time interval for each of these crops, with the rate of decrease being nearly the same from crop to crop.

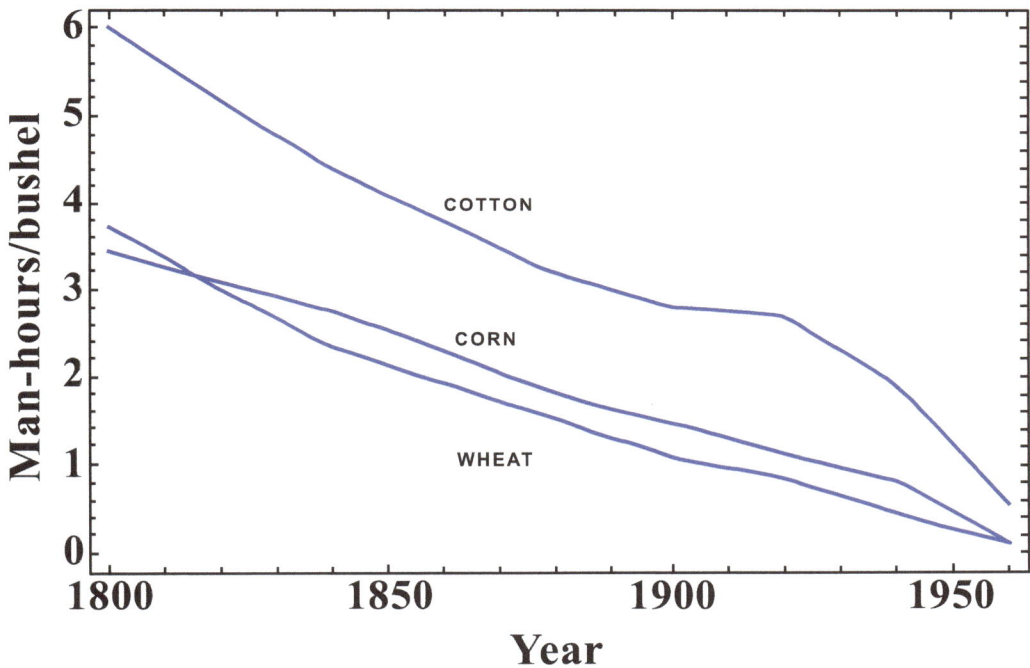

Fig. (1.3). Productivity of workers in wheat, corn and cotton culture in the United States increases with decreasing time required to produce a given bushel of produce. The data is taken from [2].

Thus, we have a fairly clear picture of population growth using the simple growth model of Malthus. This model enables us to discuss what was of importance to an agrarian society and to make predictions. But when the predictions of the growth of populations were compared with data they were not very good. When a theory does not agree with data it is the theory that must be corrected, so the next step in explaining a phenomenon is to increase the model's accuracy. Increased accuracy is achieved by including mechanisms that had been omitted from the original description and consequently complicate how the growth of population is understood. This leads to the introduction of a nonlinear term in the description of population growth.

1.2.2. Saturation

A century before Boulding posited his theorems, the King of Holland was alarmed by the essay of Malthus and feared that his small kingdom was in peril from overpopulation. In the tradition of governments large and small he commissioned a re-examination of the overpopulation question. The scientist Pierre Francois Verhulst (1804-1849) responded to the King's request, with a theory of population growth that reconciled the pessimistic view of Malthus. In 1845 Verhulst published a work noting that the growth in population was not unbounded, but was limited by a number of factors, such as food, shelter, and hygiene [10]. For example, we know that the linear increase in the food supply cannot be maintained indefinitely. There are only so many improvements in agriculture that can be made and then the food supply ceases to increase from year to year, or it increases at an ever decreasing rate until its production is constant. This is the kind of reasoning that Verhulst was able to incorporate into his mathematical theory of population growth in a direct way by modifying the rate of growth to be a decreasing function of population. The growth rate is largest at early times and monotonically goes to zero as the population approaches a saturation level.

In Fig. (**1.1**) the pure exponential population growth of Malthus is contrasted with the saturated population growth of Verhulst see Fig. (**1.4**). The linear growth in food supply is indicated to contrast how food and population change over time. Note that the two theories of population growth coincide at early times, when the level of the population is far below saturation. If there were data available on the initial growth of a population being modeled and an exponential growth law is assumed, then a growth rate could be fit and the unlimited growth articulated by Malthus and many others would be predicted. Scientists use the term 'fit' to denote techniques for determining the values of model parameters, such as a growth rate or a saturation level, from the data. Usually not all the data are used to fit the parameters. Some data are set aside so the predictions from the model with the empirical parameter values can be compared with actual data. The parameter values that give the smallest error in terms of the deviation of the prediction from the data are considered best.

However fitting the growth rate for the exponential does not necessarily require that the subsequent population growth is exponential, only that such growth is not inconsistent with the available data and the empirical growth rate. On the other hand, a more optimistic assumption would be that saturated growth can take place. Under this assumption the same initial data determines the largest population the environment can support, as well as its rate of growth. The more data available, the more accurately the potential saturation level can be determined. Ultimately it is the quality of the prediction that determines which of the two models better fits the data and of course it might be that neither is satisfactory.

Fig. (1.4). On the left is the minister Thomas Robert Malthus (1766-1834) who wrote one of the most influential books on population growth emphasizing that human misery and starvation are the only brakes on exponentially growing populations. On the right is the scientist Pierre Francois Verhulst (1804-1849), who introduced saturation into population growth and invented what later became known as the logistic equation.

Different societies at different times have adopted a number of strategies to stabilize population growth: abstention, contraception, abortion, the one-child restriction in China and perhaps even the Women Rights movement in the West. The question is whether the steady increase in resident population is due to birth alone or whether other factors such as war, economic factors, or immigration can be sufficiently large to significantly supplement the rate at which a population grows? The effects of people moving in and out of a country are not included in the Verhulst saturation model of population growth. To develop a sense for the size of the influence these external factors might have on population growth the census data for the United States from 1790 to 2000 is graphed in Fig. (**1.5**).

Raymond Pearl (1879-1940), a legendary researcher in human biology, along with Reed [11], used the Verhulst equation, also called the logistic equation, to fit the data of the population of the United States from 1790 to 1930 as shown in Fig. (**1.5**). This figure is my own fit to the data but the solid curve is essentially the same as one made by those pioneers. The population curve is projected far beyond 1930 as shown by the solid line segment and predicts a saturation level of approximately 200 million people at the new millennium. The curve fits the data extremely well for the early period and yet the prediction is obviously wrong at late times and deviates markedly from the data points after 1950 or so.

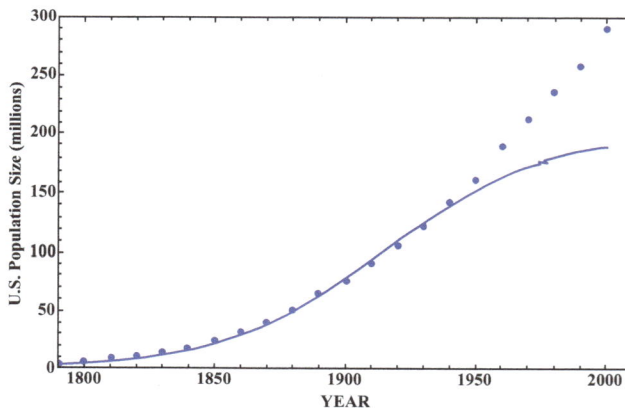

Fig. (1.5). United States population. The curve is a fit to the Verhulst equation made in the same way as done by Pearl and Reed [11] to the data from 1790 to 1930 resulting in a predicted (solid curve) saturation of approximately 200 million people.

What happened? Is the model just wrong? Or is the failure of the prediction due to the model's inability to account for other important factors? What is clear is that this simple logistic model alone is not sufficient to explain how the population data changes over time. So the intellectual map of population growth we are constructing must be clarified by looking for additional patterns within the data and from these new patterns identify explanatory mechanisms. Such prediction error prompted a number of people to suggest that the model had fundamental flaws and real populations under stress behave more like Malthus than Verhulst.

This rather arcane comparison between the two models has been well understood for over half a century, but that understanding did not prevent the re-emergence of the Malthusian view in the 1960's along with the accompanying fear of global overcrowding that it rekindled. Organizations such as the *Club of Rome* beat the drum in support of that fear under the rubric of scientific information and environmental awareness in such publications as 1972 book by Meadows *et al. The Limits to Growth* [12]. In the tradition of Malthus, their original model assumed exponential growth of world population, as well as in other indicators of modern society such as pollution, industrialization, and resource depletion. On the other side of the ledger they assumed a linear growth of the ability of technology to increase the availability of resources. As a consequence of these assumptions and the then recently available computers, using computer calculations of their equations they predicted a world-wide economic and societal collapse by the year 2000. The back cover of the book summarized their prediction:

> "Will this be the world that your grandchildren will thank you for? A world where industrial production has sunk to zero. Where population has suffered a catastrophic decline. Where the air, sea and land are polluted beyond redemption. Where civilization is a distant memory. This is the world that the computer forecasts."

However this predicted collapse did not occur and more recent computer calculations extend the predicted collapse date to sometime within the twenty-first century. This pessimistic prediction is similar to other such forecasts made in the past and that are being made even today. The key here is that the prediction is made by a computer and not by a scientist, because it is assumed that the phenomenon is much too complex to be analytically modeled, that is, no simple closed form function can be used to represent the process. The fallacy in this reliance on large-scale computer calculation is that it lacks independent checks. The simple models that check pieces of the large-scale computer calculations are dismissed, and consequently the computer output simulates reality, without providing understanding of the underlying mechanisms. Unfortunately the attitude on the part of the practitioners is often that only the computer can do the required calculations and reach the necessary conclusions, overshadowing the role of the single investigator. When the computer is used in this way, to replace rather than to supplement human reasoning, science is replaced by prophecy and the computer is the oracle.

The exponential and saturation growth models do not merely show how idealized populations change over time. They also show how scientists answer difficult questions by thinking quantitatively. My quibble with the Club of Rome is not with their politics, but with their selective use of science and the scientific method to support their preconceptions. These authors believed overpopulation is a serious problem before they did any calculations and they constructed a mathematical model that was sufficiently complicated to mimic some of reality and still prove their point. The explicit predictions they made have not occurred and after all is said and done, it is prediction that tests any scientific theory. Of course the rejoinder is that the Verhulst model did no better in predicting the saturation level in the growth of the United States population. But before I turn to this failure I should point out that Verhulst predicted the saturation population of Belgium to be 9.4 million people and their population has remained within 12% of that value since 1960.

Let us go back and look at the failure to predict the growth of the United States population a little more closely. We do this by examining a different data set from the one used to construct Fig. (**1.5**). In Fig. (**1.6**) we depict the actual United States residential population from 1970 to 1993, which is separated into native-born Americans and immigrants. These data are then used to project the total population size given by the United States Census Bureau for the period from 1994 to 2050. The projection assumes population parameters, such as, fertility, mortality, and mass immigration levels do not change very much from those of

1993. In fact, overall immigration continues to rise significantly, entailing that residential population growth will actually be higher than with the effect of immigration shown in the figure. We can see from the forecasts in Fig. (**1.6**) that the deviation from the prediction of the residential population shown in Fig. (**1.5**) can in large part be accounted for by the immigrants entering the United States and their descendants since 1970. In both Figs. (**1.5**) and (**1.6**) the deviation from the logistic equation prediction is on the order of 100 million people. The effect of continued immigration and their descendents combine with the native-born to form a linear growth of the overall population that eventually washes out the saturation growth of native-born Americans. All other suggested modifications in the Verhulst model become negligible in comparison to that of immigration.

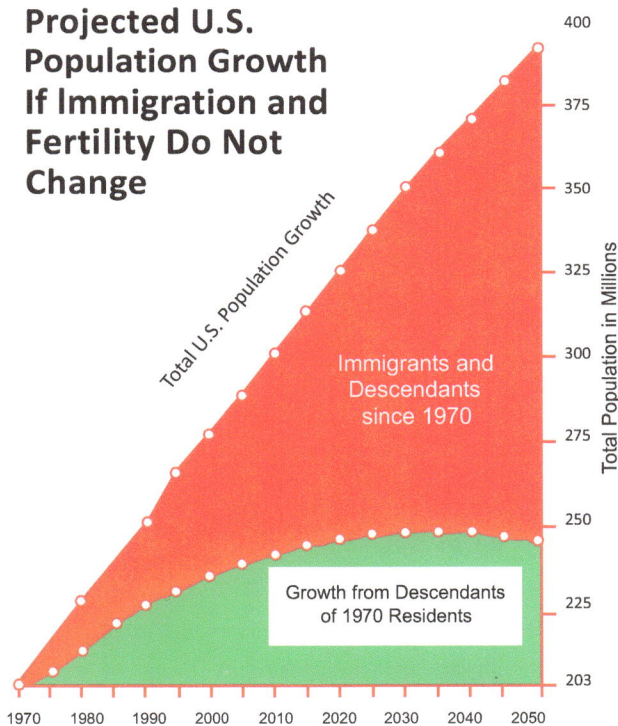

Fig. (1.6). The green lower portion of the graph represents growth from 1970 Americans and their descendants. There were 203 million people living in the U.S. in 1970. The projection of growth in 1970-stock Americans and their descendants from 1994 to 2050 is based on recent native-born fertility and mortality rates. The red upper portion of the graph represents the immigrants who have arrived, or are projected to arrive, since 1970, plus their descendents, minus deaths.

I have briefly discussed three layers of complexity that modulate understanding of how populations grow. The first mathematical model of the growth of organized society was that of Malthus in which a constant fraction of the existing population was added in each generation leading to unbounded growth. This exponential increase is the same as the fixed compound interest on a bank deposit.

The next complication was the reduction in the rate of growth with increasing population introduced by Verhulst to account for limited resources. The change in the growth rate takes into account that even a linear increase in food supply is unrealistically high.

The third complication was from the *Club of Rome* who used the Malthus model for multiple variables. This modeling recognized that even a phenomenon as well defined as population growth was not disconnected from the state of society and how society interacts with the rest of the world. Consequently the single mechanism of immigration apparently invalidates the model of Malthus, as well as that of Verhulst, when applied to individual countries.

To this picture of increasing complexity of population growth we now add a fourth feature known as 'chaos'. This last mechanism results from internal dynamics generated by the nonlinear interactions among members of the population and the subsequent instability of population growth. And of course people are not the only populations of interest.

1.2.3. Chaos

All the models introduced up to this point assume that time flows continuously like a river and population is an unbroken function of time. In principle we could know the population at any instant of time, and this is not wrong, but we do not live in such a world. In the United States a census is taken once every decade and so the population level is only known precisely, if not accurately, after each census. The continuous logistic equation has a continuous solution with the saturated form shown in Fig. (**1.5**) that smoothly connects the discrete population levels determined every ten years. This discrete set of numbers date back to the first census in the United States and because the data are discrete we might expect that a better model to explain the data would also be discrete. But here we run into an unexpected problem. The discrete form of the logistic equation that predicts population change from generation to generation has properties that the continuous logistic equation does not share. Enter nonlinear dynamics and chaos theory.

In the 1970s it became apparent to a handful of scientists that nonlinear dynamics offered a new way of understanding complex phenomena and a number of us banded together to form The La Jolla Institute in 1976. The research core of this non-profit organization consisted of Elliott Montroll from the University of Rochester, Kenneth Watson from UC Berkeley, Irwin Oppenheim from MIT and myself the corporate research memory. Shortly after its formation an opportunity presented itself for The La Jolla Institute to extend the influence of this new field of study into the physics mainstream. In 1978 the National Science Foundation solicited proposals to form an Institute of Theoretical Physics. Montroll was convinced that a proposal emphasizing the importance of nonlinear dynamics in the future of physics would be positively received, so he and I wrote and submitted a La Jolla Institute proposal. We proposed the formation of a Center for Studies of Nonlinear Dynamics in which the fundamental problems associated with the mathematics of complex physical phenomena would be studied. Moreover this Center would be a division of The La Jolla Institute. Our proposal was among the four finalists in contention for the new physics institute, but it was eventually awarded to UC Santa Barbara. The founding director of the new institute Professor Walter Kohn had in fact been on the La Jolla Institute proposal being part of a collaboration with UC San Diego, where Professor Kohn was at the time.

After the award of the National Science Foundation grant to UC Santa Barbara, Kenneth Watson and I talked about what we might do with this apparently excellent proposal for a new nonlinear dynamics research center. In the private sector research is rarely shelved, but is tweaked, polished, extended and recycled; so are the proposals. At that time I had been doing research with Watson for a few years on the properties of wind generated water waves on the ocean surface. He was an expert in physical oceanography and had been doing research for the Navy for a long time. He explained to me that all the fundamental research problems of interest to the Navy involved fluid dynamics, which was a notoriously difficult area of application of nonlinear dynamics. Subsequently we rewrote the proposal putting a decidedly fluid dynamic spin on the proposed research and submitted it to the Office of Naval Research. After a number of trips to Washington DC and a number of presentations to General Officers, it was funded at a level of $500K per year. An impressive sum for the 1970s.

The Center for Studies of Nonlinear Dynamics was established in 1979 and was the first Center of its kind in the world, being the harbinger of the research explosion into nonlinear dynamics and complex phenomena in the 1980s and 1990s and extending in multiple guises into the twenty-first century. The research orientation was physical oceanography, as we had promised the Navy, but the mathematics was fundamental and the Center attracted the top scientists in the world and their newly minted doctoral students to come, relax and work with the research staff on the future of nonlinear dynamics. The strength of the Center lay in the approximately half-dozen or so post doctoral researchers that came for a two or three year stay, to learn about the application areas of nonlinear dynamics. One such researcher was my friend Michael Shlesinger, a student of Elliott Montroll, who went on to become the first program manager for nonlinear dynamics at the Office of Naval Research.

One visitor to the Center was Robert May, who was perhaps the first scientist studying population dynamics to take the mathematical properties of discrete nonlinear equations seriously [13]. He studied the behavior of the discrete logistic equation, which is a difference equation, since its solution is determined by the difference in population levels between successive generations. But his interest was not so much in the mathematics of the difference equations as it was in their curious manifestation in the real world [13]:

> "Not only in research, but also in the everyday world of politics and economics, we would all be better off if more people realized that simple systems do not necessarily possess simple dynamic properties."

He was referring to the failure of nonlinear difference equations to have predictable outcomes. Such equations are unpredictable in the sense that extremely small changes in how the system begins to change, can have dramatic effects on where the system ultimately ends up. It is as if every nonlinear system is at a tipping point and the smallest change in the initial stimulus can send it over the edge. In the case of chaos however there are multiple tipping points leading to a wide variety of unexpected final states [13]:

"...that the apparently random fluctuations in census data for an animal population need not necessarily betoken either the vagaries of an unpredictable environment or sampling errors: they may simply derive from a rigidly deterministic population growth relationship such as the logistic equation."

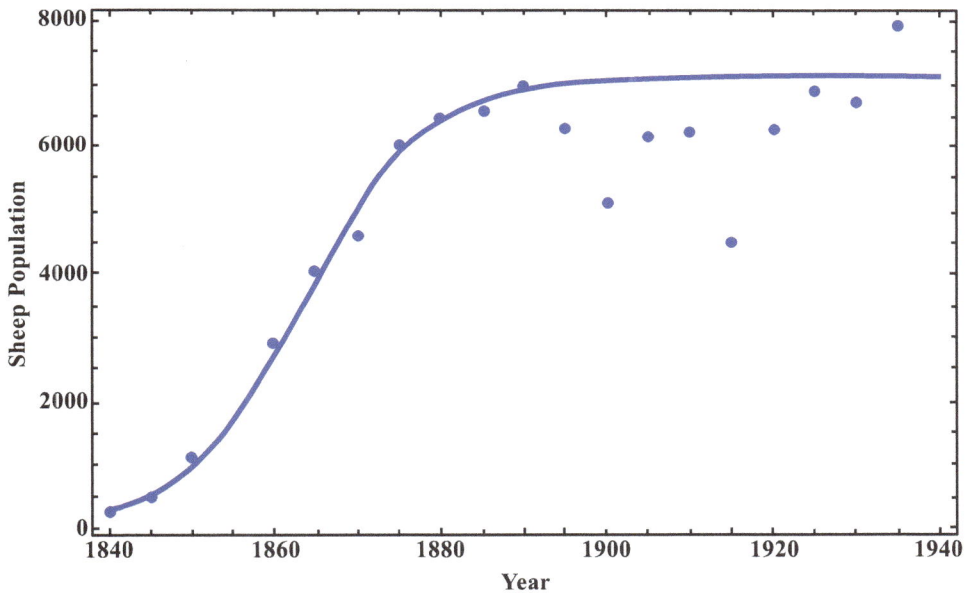

Fig. (1.7). The size of the sheep population is Southern Australia between the years 1838 and 1940 averaged over five year intervals is given by the dots. The solid curve is the best fit to the data using the logistic (Verhulst) equation. [Adapted from Davidson [14]].

An example of May's concerns is given by the graph of the sheep population in Southern Australia depicted in Fig. (**1.7**) over the century beginning in 1838. It is evident that the logistic or Verhulst equation coincides with the early growth of the sheep population, but at the beginning of the twentieth century erratic fluctuations seem to take over. Of course the variability of such things as the availability of pasture lands and rainfall contributed to this randomness, but May cautions that a significant contributor to such fluctuations may be intrinsic to the nonlinear dynamics of population growth. Consequently intervention into the growth process to reduce fluctuations to make growth more predictable must be guided by an understanding of what humans can influence and what they cannot. This sage advice has application far beyond the growth in sheep populations.

Billions of tons

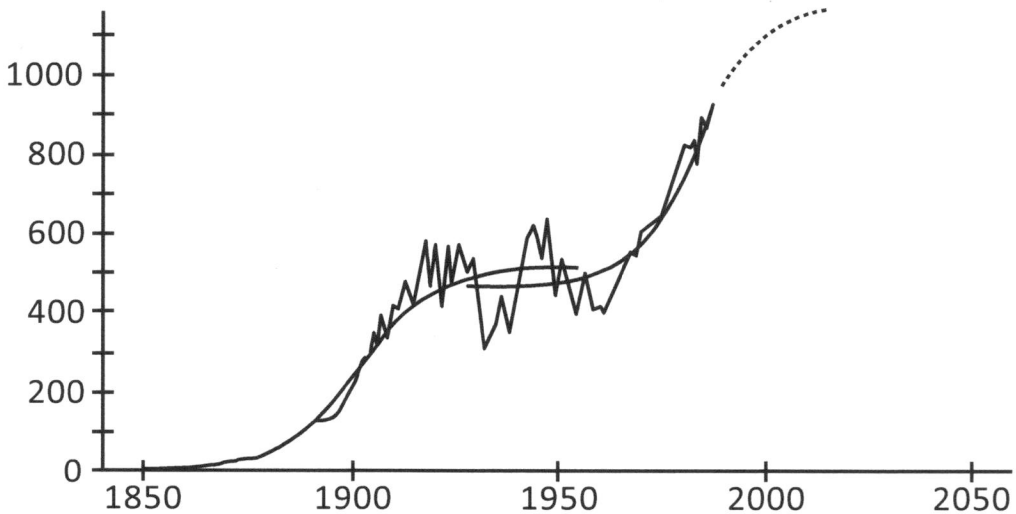

Fig. (1.8). Annual production of bituminous coal in the United States. The sigmoidal curves shown are logistic fits to the data for the respective historical periods. The interim period shows large fluctuations of a chaotic nature [15]. The small circles show what happened during the twelve years following the original prediction [16]. [Private communication of permission].

A property of a logistic curve is that it starts out with an increasing slope in the exponential region of Malthus and ends with a decreasing slope in the saturation region of Verhulst. The slope changes from increasing to decreasing at the inflection point at the center of the curve. It is at the inflection point that a number of data sets show an oscillating fluctuation. These erratic fluctuations have been interpreted as being chaotic and a significant number of non-biological growth phenomena display this behavior. In Fig. (**1.8**) the annual U.S. production of bituminous coal is depicted [15]. In this figure the fluctuations in the coal production between 1920 and 1960 could belong to both the saturation region of the first logistic curve to the left of 1925 as well as to the beginning of a second logistic curve to the right of 1925. The solid curve is the data up to 1992 the open circles are the data in the subsequent ten years and agree with the logistics curves fit to the earlier data [16].

The last example depicted in Fig. (**1.8**) shows that the growth of things other than biological populations can be described by the logistic equation. This kind of multiple logistic growth curve is further discussed in the next section. We now turn our attention to the growth of various technologies in society and finish the sketch of our deterministic mental map of the world.

1.3. SOCIOTECHNOLOGY GROWTH AND MEASURED THOUGHT

The all too brief discussion of population dynamics was presented, in part, to show the often catastrophic effects of even the simplest assumptions on how the world operates. An assumption can appear benign even when it is not, but it is always linked to a fundamental belief we hold about the world. In our technological society many of our fundamental beliefs are the result of the complexity we experience on a daily basis. For example, in the latter part of the seventies I considered buying a home computer, an Apple II⁺. At the time that would have been a major investment, so I brought the matter up with my wife; who is an artist and normally a very reasonable person. She saw the purchase as an extravagance, right up there with a new house or a luxury car. I waxed poetic on how the computer was going to change everything we knew and having one in the house would give our sons a head start in education, but she was adamant. Remember that, at the time, the Internet did not exist, much less *Google*, and the personal computer had not yet transformed society. My final argument was that the computer would keep our sons from becoming bored with school, a problem that both of us certainly struggled with growing up. My wife could not see the computer's value to us and countered every argument I offered. In retrospect I realize that I had not anticipated most of the real applications to which our sons would eventually put the computer. But even with my limited understanding I was convinced that we should buy a computer and so we continued talking.

In the spirit of compromise, since we had reached an impasse, we decided to resolve the problem by seeking advise from a senior scientist and friend who was a Physics Professor at UC Berkeley. At a La Jolla Institute office party a day or so after the computer question came up my wife asked Ken Watson about the educational value of computers and whether our oldest son who was eight could use one. His response was: "Can he play tick-tack-toe?" We bought the computer.

The personal computer was certainly a disruptive technology that changed the way our family developed. Such disruptive technologies tend to do that, they change the way we view the world, in ways we cannot anticipate. Just ask my wife. For me it is perhaps more obvious. I now have a meeting every Friday morning with a group of international scientists on Skype connecting good friends and scientists in Italy, Germany, Texas and North Carolina. Such a thing would have been literally incomprehensible to me as a graduate student. My younger son, the one in physics, does this as a matter of course. I doubt whether he even thinks about how this evolution in technology changed the form of his experience of graduate school from my own. But then he has no particular reason to think about that comparison.

1.3.1. Evolution as Replacement

Visualize an evolving technology (and society) as a sequence of replacements, where existing devices and applications are being continually replaced by those that are technically superior. Elliott Montroll observed that a remarkable feature of a replacement technology is its autocatalytic nature, each innovation catalyzing the generation of its own replacement. Here the technological and social evolution is a consequence of a sequence of replacements of one technique, Or idea, tradition, or artifact, by another [2]:

> " This statement is in the Darwinian spirit of survival of the fittest, with each new mutation or species struggling to find its niche, sometimes at the expense of displacing or replacing the older forms. Once a virile mutant or new form established itself, it would be expected to propagate, continuing to replace its competitors until it reached an equilibrium saturation level."

Consider how one technology makes another obsolete and replaces it in society. At one time, not so long ago, we would watch movies on television by playing a videotape. We no longer play tapes, since the transfer from VHS to DVD is almost complete and nearly everything we might want to see is on DVD. Our children probably do not remember VHS and if they do they would think that such arcane technology are in the category of black and white films. How did this replacement of VHS by DVD come about and was it predicted?

Montroll formulated a law of social dynamics based on the idea of a sequence of replacements. The notion of evolution under this law exceeds the meaning of Darwin, in that it includes changes in social networks over time, changes in products, as well as, improvements in modes of transportation. In the same way the meaning of the phrase 'sequence of replacements' is flexible as well and adapts to what is being investigated: mutations in biology; chip capability in computer technology; the train replacing the stagecoach and in turn its being replaced by the airplane in transportation.

The social law states that in the absence of any "force", whether social, economic, or ecological, the population growth law is that of Malthus. This is the exponential growth of the dynamic variable when there is no competition. The computer industry provides one of the clearest examples of this law. In 1965 Gorden E. Moore, the cofounder of the computer-chip giant Intel, noted that the number of transistors on a computer chip seemed to be doubling approximately every eighteen months. This became known as Moore's law and as shown in Fig. (**1.9**), it is one of those uncanny patterns in complex social phenomena that enable prediction without real understanding. In four decades the increase in the number of transistors on a chip has increased by a factor of a million in the same manner as those pennies you were going to pay me. What made this increase possible were the advances in technology that enabled making transistors smaller and smaller. But there is always a limit to such growth and there is apparently no more room at the bottom. So we must look for either replacement technologies for the transistor or competitive processes that disallow making the transistors any smaller. Heat seems to be the limiting physical process that transistors have come up against in 2011 and dissipating this heat is the technical challenge inhibiting further size reduction. It is interesting that a paper in the premier journal *Science* in January 2012 reported research on a new nanoscale architecture that provides one way to circumvent the exponential increase in heat generation at these extremely small scales.

It is now possible to define a social force as anything that causes the social law to be violated, such as the disruption of Moore's law. This definition of a social force is consistent with the clever way Sir Isaac Newton defined a mechanical force in physics as anything that violated the law of inertia. The social force with the simplest mathematical form is one that is linear in the population level. Consequently, we obtain the form suggested by Verhulst to modify the growth equation as the simplest form for a social force that induces saturation.

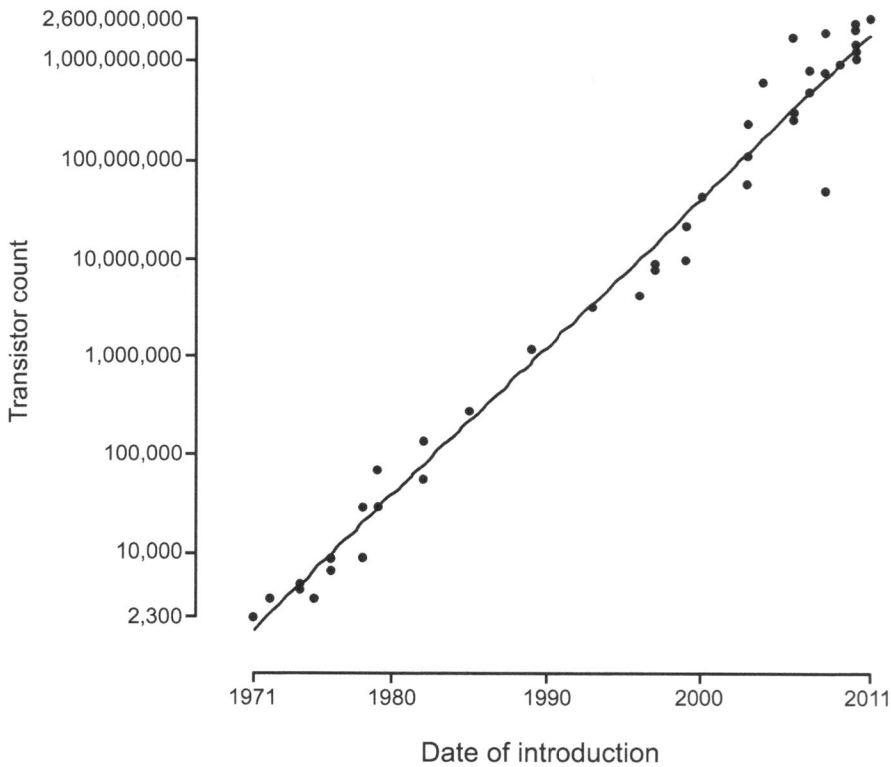

Fig. (1.9). Plot of CPU transistor counts against dates of introduction with the identifying label of the specific chip deleted. Note the logarithmic vertical scale; the straight line fitted to the data corresponds to exponential growth, with transistor count doubling every two years. [Adapted from http://en.wikipedia.org /wiki/Moore's_-law].

The logistic equation has been used to model social (product replacement) phenomena as well as population dynamics. Product replacement associates the dependent variable with the fraction of the available market attained by a particular product. The solution to the logistic equation describes how one product replaces another over time. In Fig. (**1.10**) the ratio of the fraction of the market captured by a new product to that remaining for the old is plotted on semi-logarithm graph paper as a function of time. Data that lie along a straight line are consistent with a solution to the logistic equation.

The evolutionary curves shown in Fig. (**1.10**) are fitted to data using the logistic equation by Fisher and Pry [17] for several industrial replacements. They show that if a new process or product excites trade sufficiently to absorb 10-15% of the market, it is highly likely that it will win an increasingly larger portion until it completely dominates the market or until its own new replacement appears. The product replacement of soap by detergent in the United States and Japan is truly remarkable. In the United States detergent virtually takes over the market in fifteen years and in Japan it is accomplished in a decade. The time for open hearth processing to replace Bessemer processing of steel required sixty years. The replacement time is tied to how the product or process is embedded within the society and consequently how long it takes society to adapt to the new technology.

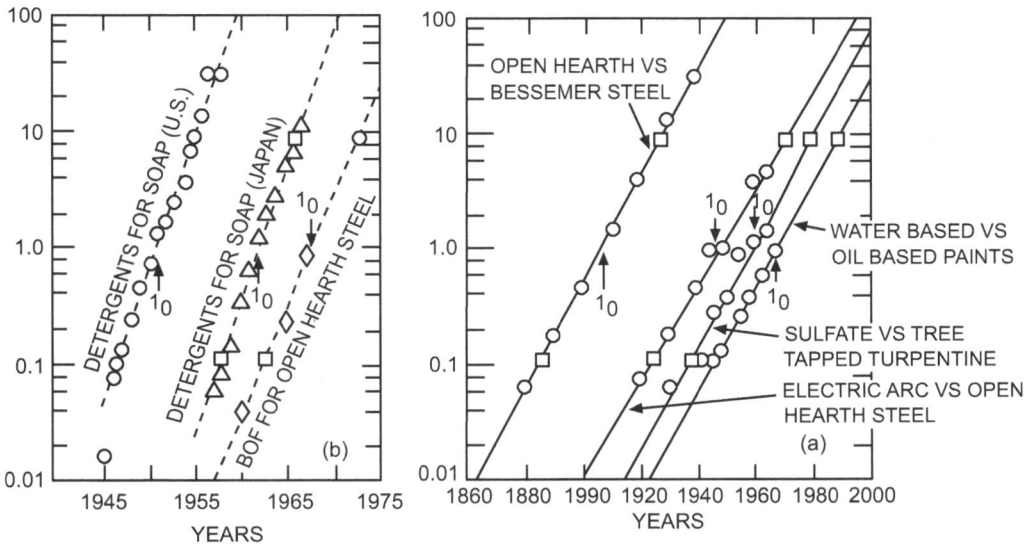

Fig. (1.10). Substitution data are fit to the Fisher-Pry model [17] (logistic equation) for a number of products and processes: all data are for the United States except where indicated.

Some might criticize Fig. (**1.10**) because the citation is forty years old. They might ask whether the Fisher-Pry model has successfully predicted the market takeover of more recent technologies. In fact the model has withstood the test of time in the technological forecasting community. In Fig. (**1.11**) the data from a number of replacement technologies are shown along with the solutions to the Fisher-Pry logistic equation. The solid saturation curves are the percentage of a new technology in the market. Note that this is a different representation from that in Fig. (**1.10**) and I use it here only because it allows for a visual comparison with the biological population growth model of Verhulst.

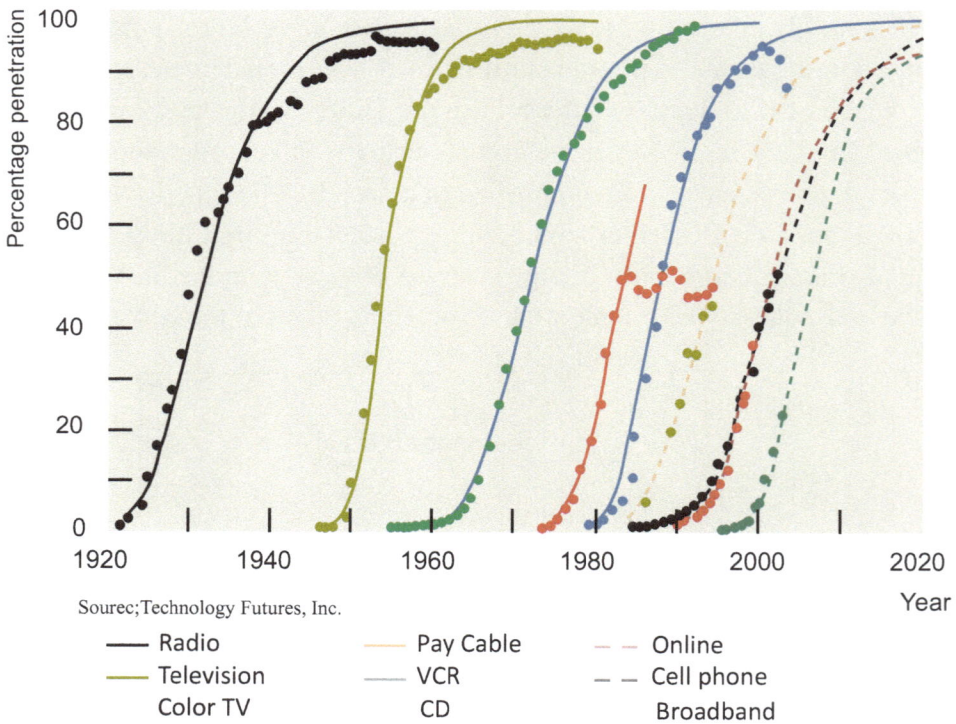

Fig. (1.11). The dots are data points for each of the indicated replacement technologies [18]; from left to right, diesel locomotives, basic oxygen and electrical steel, front disk brakes, SPC switching, office PCs and local area networks. The solid curves are solutions to the logistic equation of the Fisher-Pry model.

It is also possible to discuss the mode of transportation introduced previously as a way to measure affluence in terms of technology replacement. Canals were replaced by railroads; the railroads ceded to an intercoastal highway system; the highway transport system was in its turn replaced by international airfreight. The time between the 50% point of each of these new modes of transport is approximately the same, that being, 55 years, see Fig. (**1.12**). It is also possible to stack the modes of transport to show how the efficiency in the new modes drives out or replaces the existing mode. This sequential replacement interpreted in terms of multiple saturation is explained in the next subsection.

1.3.2. Multiple Saturations

We are now in the same position for understanding technological growth using the product replacement model that we were in understanding population growth when we ended the last section. We have transitioned from a simple exponential growth of the process in isolation to the saturation growth induced by complex market forces. This relatively simple picture captures the saturation pattern seen in the data and explains the effect in terms of two heuristic parameters, the rate of growth and the saturation level.

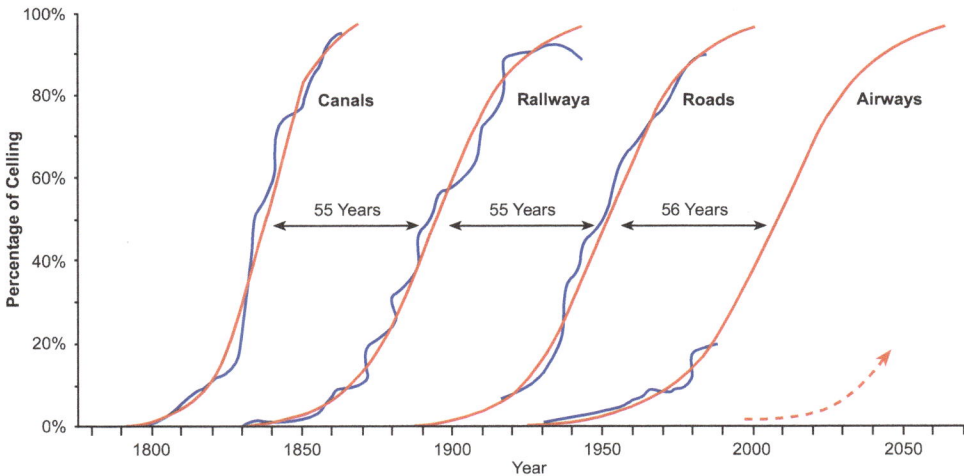

Fig. (1.12). The change in the percentage of market share held by a newly introduced modes of transport as a function of time. The distance between inflection points of the sequential modes appears to be constant.

It probably bears mentioning that the term disruptive technology in addition to being an accurate description of certain innovations that completely change the social environment into which they are introduced, also has a more technical meaning. The technical meaning of a disruptive technology was introduced by

Christensen [19] in order to discuss a certain kind of innovation that had no customer base when introduced and therefore had no market rationale. In spite of this inauspicious beginning such a technology goes on to completely dominate a particular segment of the marketplace. This is very different from a succession of product replacements.

We now have a clearer picture of how a new technology grows to saturation as it penetrates the market and eventually completely replaces the old technology. But what happens when this growing technology is faced with its own replacement? An exemplar of this type of growth using the rate of discovery of chemical elements was given by Derek de Sola Price [20] and that data is plotted in Fig. (**1.13**). From this figure we see that there is a sequence of logistic curves stitched together; as an old technology reaches the saturation level a new technology emerges to take its place. In the case of the discovery of new chemical elements each burst of discovery was predicated on the invention of a new technology necessary for its measurement.

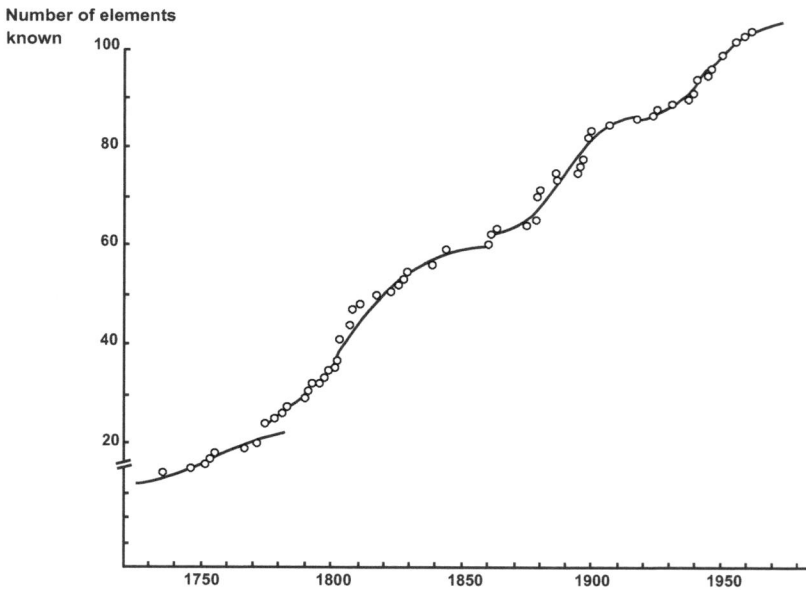

Fig. (1.13). Number of chemical elements known as a function of date. After the work of Davy there is a clear logistic decline followed by a set of escalations corresponding to the discovery of elements by techniques that are predominantly physical. Around 1950 is the latest escalation produced by the manufacture of trans-uranic elements [20].

At the intercept of the slow growth in the old technology with the growth spurt in the new replacement technology two effects have been seen in the data. The first had to do with the instability of the nonlinear dynamics and produced chaotic variations in the growth rate, see Fig. (**1.8**). The second effect is the sequence of smooth connections between logistic curves as depicted in Figs. (**1.13**) and (**1.14**). In the latter figure the number of known animal species are plotted in the same way as the number of elements in the former figure. It is clear that the overall logistic curve can be decomposed into a cascade of smaller logistic curves, which suggests a dynamic self-similarity in the underlying dynamics. The notion of self-similarity is a fundamental concept for the understanding of complexity and it is evident that it involves the repetition of a form or property at multiple scales. This is taken up in Chapter Three and its implications for the physical, social and life sciences are discussed. The changing environment over time is manifest through the sequence of interconnected logistic curves. In this way the maximum rate of growth of each of the subprocesses are the same. Such a rule is defensible in terms of evolution because it reflects the stability of resource allocation for each species in Fig. (**1.14**).

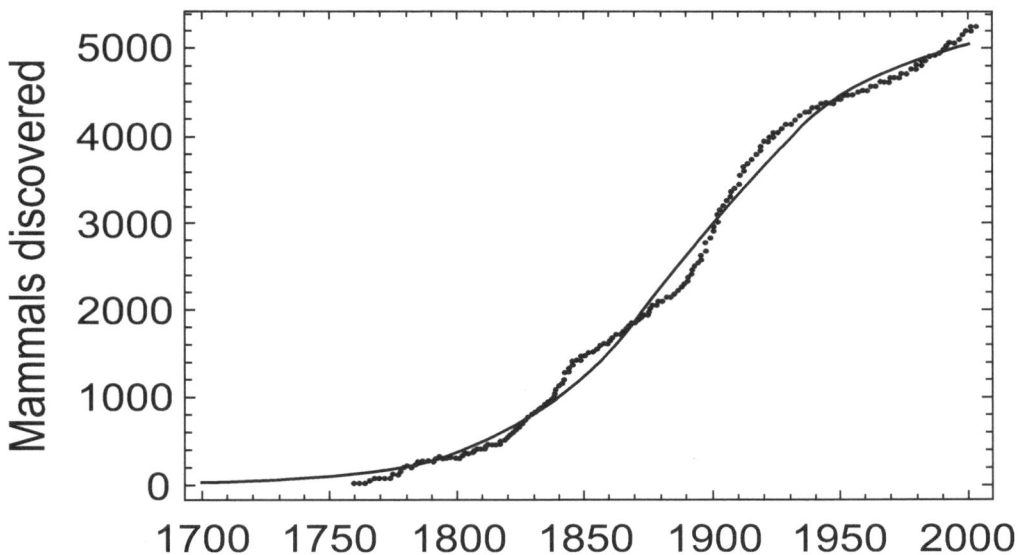

Fig. (1.14). Cumulative number of mammalian species discovered over time from 1760 to 2003. [From [22] with permission].

In Fig. (**1.14**) the cumulative number of mammalian species discovered over time from 1760 to 2003 is depicted. A single logistic curve is used to fit the overall data, but it is clear that the data can also be interpreted in terms of an interconnected sequence of logistic subprocesses [21]. This arrangement in terms of logistic subprocesses has more to do with the environment and the development of methods to discover new species than they do with the species themselves.

1.4. ASSESSMENT

What I have discussed in this first chapter are ways in which to replace a myriad of events with simple models that are intended to capture the dominant features of those events. If the world did not change over time I could add more and more detail with each repetition of an event and eventually have an accurate reconstruction of a successful economic relationship, of a nurturing family, or of a supportive organization. But things do change even if my reactions to them do not. Human reactions resist change because people react to how they believe things ought to be, rather than how they are. If my reactions are sufficiently inappropriate I slightly adjust my mental picture so that the next time I am called to respond I will be ready. Such adjustments become increasingly hard to make with age and bedrock beliefs are almost never modified.

Malthus was a minister whose mental map depicted the human condition as miserable and society as being fundamentally flawed. It is not surprising that his quantitative model supported this perspective. Verhulst, on the other hand, was a scientist and although sharing many of the concerns of Malthus developed a quantitative model that through human intervention and control was able to mitigate much of the misery anticipated by Malthus. The model of Verhulst was even able to provide useful quantitative predictions of the population level for closed societies, such as Belgium and helped identify powerful growth mechanisms for open societies such as immigration to the United States. It is interesting that Verhulst did not believe that social decisions should be based on simple mathematical models, whereas Malthus did.

One such belief that I come back to again and again is whether our world is fair, or more accurately whether society is fair and what, if anything, can be done about it, if it is not. What is the evidence regarding fairness one way or the other? Is the wealth of the country fairly distributed? Is the judicial system fair and unbiased in its dispensation of justice? Or perhaps of equal importance is whether people are predisposed to believing that the world ought to be fair, even if it is not? If the latter is true do people change their expectations, or do they change the world?

Some ideas people steadfastly believe to be true, but as Leo Tolstoy (1828 - 1910) put so eloquently:

> "I know that most men, including those at ease with the problems of the greatest complexity, can seldom accept even the simplest and most obvious truth if it be such as would oblige them to admit the falsity of conclusions which they have delighted in explaining to colleagues, which they have proudly taught to others, and which they have woven, thread by thread, into the fabric of their lives."

In subsequent chapters I will explore ideas that may cause some to re-examine their most deeply held beliefs. The examination process can be uncomfortable, not because we have not thought about these things before, but because we settled them in the distant past. Carefully examining beliefs and consolidating what is held true about the nature of society was part of the rite of passage at the university and should not disrupt a comfortable middle life. On the other hand such disruption may enable us to address those vague feelings of apprehension that occasionally overshadow the voices of politicians and experts that explain to us how the world will be in the future, if only we do what they ask.

So let me begin by pulling on one of those threads from Tolstoy's tapestry. I identify other fibers in subsequent chapters, but the thread from this first chapter is woven into the foundation of western civilization. That is understanding society through science and the use of quantitative models. In subsequent chapters I pull on this strand and others in an attempt to determine what unravels, but I will also try to mend the rift by spinning a new yarn.

CHAPTER 2

Uncertain: A Simple World View

Abstract: Everyone knows the future cannot be predicted and yet fortune cookies are invariably received with pleasant anticipation. In this chapter we review how science came to terms with uncertainty, through the invention of statistics and probability, but perhaps more importantly, how this world view was made compatible with the clockwork universe of Newton. If the changing events of one's life are treated as being linear, then response is proportional to stimulus, with perhaps a little error. But the error in this view is subject to law, and is therefore controllable. The linear world view, with Normal statistics to explain uncertainty, is the model of reality adopted by most people, either implicitly or explicitly. It is this world view that promotes the idea that equality and fairness are not only what is true, but more importantly they are what ought to be true.

Keywords: Adrian, Drunkard's walk, Gauss, Handicapping, Linear, Medicine, Normal statistics, Prediction, Probability, Psychophysics, Scaling, Simple models, Sociophysics, Uncertainty.

At the opening of the nineteenth century a new scientific view of the world, based on statistics, was introduced by the German polymath Karl Fredrich Gauss (1777-1855) [23]. In the same year an identical view was independently published across the ocean by the American mathematician Robert Adrian (1775-1843) [24]. These mathematicians solved a great mystery that had confounded scientists since the acceptance of Sir Francis Bacon's (1561-1626) assertion that the best way to answer questions was through experiment. Bacon is credited with fathering the scientific method; an inductive logical procedure for isolating the cause of a phenomenon through the judicious use of experiment. For example, humans were curious about the nature of lightning since the beginning of recorded time, but it remained for Benjamin Franklin (1706-1790) to initiate experiments on lightning

Bruce J. West

in 1749, see Fig. (**2.1**). As pointed out in the Encyclopaedia Britannica Franklin's deductions based on his carefully designed experiments remained the best information on lightning for 150 years. In spite of their quality Franklin's experiments on electricity suffered from small uncontrollable variations in the results; a vexing situation common to all other experimenters of the time.

Fig. (2.1). Sir Francis Bacon (1561-1626) the inventor of the scientific method is depicted on the left. Benjamin Franklin (1706-1790) used Bacon's method with considerable success.

Gauss and Adrian were the first to explain why the vacillation in results from experiment to experiment occur, and never produces the same value of a given variable twice. This academic discussion did not influence the horizons of the farmer planting his crops, nor did it change the vision of the landlord over- seeing his holdings, but in the cities where innovation was flourishing the intelligentsia was listening. What captured the imagination of the nineteenth century philosophers and scientists (natural philosophers) was that unpredictable random variations obey a law in the same way that predictable physical phenomena obey laws. What is unique and unpredictable is not completely arbitrary. The law of randomness was expressed through the interpretation of the bell-shaped curve of Normal statistics depicted in Fig. (**2.2**). This curve peaks at the center where the largest number of experimental values occurs. Most results are in the immediate vicinity of the mean, so that the center contains the greatest concentration of results. The greater the distance a value is from the peak the less often it is observed in the experimental data.

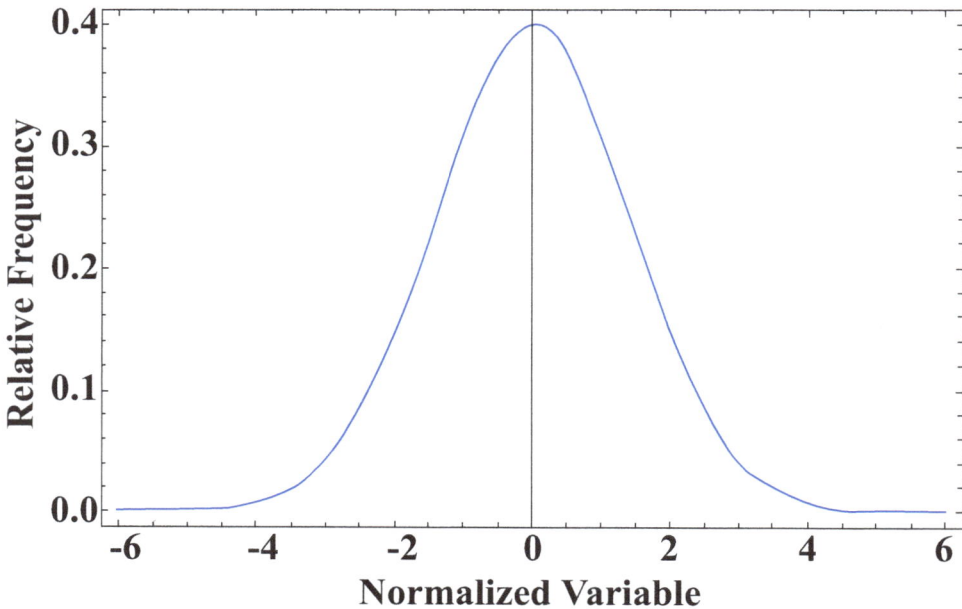

Fig. (2.2). The bell-shape curve of Gauss, Adrian and Laplace concerns errors. Consequently the average of the experimental data is subtracted from each data point so the curve peaks at zero, and the data are divided by the standard deviation of the data (width of the distribution) so that the normalized variable is dimensionless. The peak occurs at the point of zero error, that is, the average of the data. The region two standard deviations above and below the average value contains 95% of the errors (the deviations from the average value).

The bell-shaped distribution was interpreted as the law of frequency of error, since it was believed that measurements ought to have a correct value; one fixed by the underlying dynamics of the phenomenon being investigated. For convenience let us agree to refer to this as the law of error. This view of physical phenomena was and remains consistent with Newtonian mechanics that determines the ballistics of cannon and rifles, the inertia of horse and carriage, and the orbits of the planets. The world was understood to be a clockwork mechanism and therefore variables ought to be quantifiable and predictable, even those aspects pertaining to the so-called soft sciences, such as psychology and sociology. However the complexity of the world manifests itself in the measurements not being as predictable as those described by Newton's laws.

2.1. NORMALCY

Scientists in the eighteenth and nineteenth centuries, such as Franklin, expected that given the certainty of the equations of physics they would be able to accurately predict the outcome of well-designed experiments. They were therefore consistently surprised when no two experimental outcomes were exactly the same and could not be made the same regardless of the experimentalist's skill, the quality of the instruments, or how well the experiments were designed. The measured values were rarely identical to the predicted one and were randomly scattered in the vicinity of the prediction. Gauss and Adrian hypothesized that the measured values cluster around the prediction, where the bell-shaped curve peaks, with as many data points below as there are above this value. The bell-shaped distribution is symmetric about the average value, forcing Gauss and Adrian to independently conclude that the mean is the best representation of a given collection of experimental measurements.

This information is compacted into the universal curve depicted in Fig. (**2.2**). Of course, the distribution curve did not start out as universal, but depended on whether the measurements being investigated are the heights of co-eds on a college campus, the spacing between cars in moving traffic pattern, the variability in the measurement of the electric current passing through a resistor and so on. Each distribution has a different average value and a different width that depends on the units of the process being measured. Consequently, the procedure is to subtract the average value from each data point and then divide the resulting data by the width of the distribution (the standard deviation). In this way the universal bell-shaped curve peaks at zero (the average has been removed) and the width of the universal distribution is unity (the units of the data have been divided out). In terms of this normalized variable the law of error is universal.

Gauss and Adrian postulated that the deviations from the mean be interpreted as measurement errors. The deviations are assumed to be errors because the mean is thought to be the correct value of the variable – the predicted value that the experiment ought to yield, each and every time. Thus, even when statistics enter the picture, where uncertainly blurs what is expected, the scientist still believes there ought to be a proper value that characterizes the process. In this interpretation statistical fluctuations do not invalidate the mechanical world view; rather they complement it, making predictions only slightly less certain than the ticking clock. The universal distribution peaks at zero error, corresponding to the 'correct' average value, and consequently this is the most likely outcome of an experiment.

Here errors are not a consequence of false reasoning, rather they have to do with the precision of numbers and the results of experiment. Errors are of two kinds. They can be either random and associated with the experimenter, or they can be systematic, for example those associated with the measuring instrument. The instrumental errors may be made smaller and smaller and in principle eliminated from the observations by improving instrumentation. The experimenter's errors are uncontrollable and unpredictable influences that occur during measurement and constitute a fundamental uncertainty in the data record that we are stuck with.

Shortly after Gauss and Adrian published their groundbreaking research the Marquis Pierre-Simon de Laplace (1740-1827) presented a proof of the much celebrated central limit theorem and established that the validity and domain of application of the Normal distribution is much broader than foreseen in the law of error, see Fig. (**2.3**). Mark Kac, a well-known mathematical physicist, used to say about equations in his lectures that a demonstration is all that is required to convince a reasonable person, but a mathematician requires a proof. Consequently the physical sciences are filled with demonstrations made by reasonable scientists and mathematics is filled with proofs constructed by unreasonable mathematicians. However, scientists take solace, if not refuge, in the fact that such mathematical proofs exist and they often discuss the properties of complex phenomena, using the conditions required by such proofs, as if they were also true in the phenomenon being explained. Such assumptions are often shown in retrospect to be unjustified, but they are often useful in the short term.

There are four conditions that have been used historically to prove the central limit theorem and that can be simply expressed in the language of the law of errors as: 1) the measurement errors are statistically independent; 2) the errors are additive; 3) the statistics of each error is the same and 4) the width of the distribution of errors is finite. These four assumptions were either explicitly or implicitly made by Laplace, Gauss, and Adrian.

Fig. (2.3). On the left is Karl Fredrich Gauss (1777-1855) one of the inventors of the Normal distribution and the law of error. On the right is Marquis Pierre Simon de Laplace (1749-1827) who provided the first proof of the central limit theorem, which is the general mathematical basis for Normal statistics.

2.1.1. Examples

Everyone, from those in Third World countries to those in the industrialized West, assuming the latter occasionally get out of the city, have gazed up at the night sky and observed the twinkling of stars. The physical reason for this apparently small erratic movement lay not in the stars, but in the atmosphere, which is not stationary, but flows as a multi-layer fluid in parallel with the earth's surface. Some layers of air move in smooth laminar flow, as we proceed upward from the irregular surface of the earth. Other layers move in turbulent bursts and like the flow of water around rocks, tumble in erratic patches. If the motion of the atmosphere were entirely laminar the stars would be seen as unvarying points of light in a velvet background and their location would be easily determined.

However atmospheric turbulence produces slight changes in the direction of propagation of the starlight causing apparent shifts in the position of the source of the light. If two adjacent atmospheric layers are separately turbulent the direction of the starlight is randomly changed twice, with the second change being independent of the first. The changes in starlight direction are additive from one fluid layer to the next and since each random change has the same physical cause the statistics of the changes in direction are also the same. Finally due to the shortness of the time the light interacts with the atmosphere's layers the angular change in direction of the light is small. These conditions of independence, additivity, having the same statistics and small amplitudes are all necessary to satisfy the central limit theorem. Consequently one expects that the changes in the apparent position of a twinkling star is a Normal distribution peaking on the actual location of the star in space. This is in fact what was found. Gauss solved the problem of determining the location of a twinkling star and in doing so invented the Normal distribution which often bears his name in the physical sciences.

The mathematician Poincaré (1854-1912) commenting on the Normal distribution in *The Foundations of Science* [25] said:

> "All the world believes it firmly, because the mathematicians imagine that it is a fact of observation and the observers that it is a theorem of mathematics."

Such remarks were a caution that phenomena do not always satisfy the mathematical conditions under which a theorem is proved. However by the time Poincaré and others were issuing such warnings the Normal distribution had been accepted by scientists in many fields, most of whom adopted the intuition of Normalcy and ignored the cautionary tales of mathematicians.

Perhaps a more intuitive example of the application of the central limit theorem comes from outside the physical sciences. Consider the daily variation in the profit of a particular stock in the stock market; the profit is quantified by the percentage change in the price of a stock over a specified period of time. If my stockbroker forecasts that the stock market will rise, and it subsequently goes up, I attribute my good fortune to his understanding of the market and his ability to make predictions. He may have great success with my investments, but this does not imply that she has a scientific understanding of the behavior of the stock market. The dominant view among market professionals is that the erratic motion of a stock is predominantly a random process; one driven by external events (coupling to the environment). The events can be traced to national and international issues related to peace and war, broad economic successes and failures of smaller segments of the world economy, crop failures, shifts in local political, bank and mortgage interest rates, population trends and much more. These "causes" tend to mask underlying market mechanisms that might be characteristic of a 'social or economic law' or reflect the intrinsic value of a company whose stock is being discussed. A particular broker may be successful in this random environment, but for each one that succeeds, there are hundreds that do not.

Cootner compiled a number of important papers in his book *The Random Character of Stock Market Prices* [26] starting with a certain model of market behavior called a drunkard's walk. The model is that of a person who is so drunk that they lose orientation and randomly change direction with each step, producing an erratic path that is not too different from the starlight passing through the turbulently layered atmosphere. The model is able to determine the probability that the walker is a certain distance from her starting point after a given number of steps. The application to the stock profits interprets each of the daily changes in profit as a step in the walk and the distance travelled, after n steps as being the total profit after n days, which could be positive, negative or zero. A modern description of this process is given in any of the ten printings and multiple revisions of Malkiel's remarkable book *A Random Walk Down Wall Street* [27].

2.1.2. Drunkard's Walk

At the turn of the twentieth century Bachelier [28], a student of Poincaré, argued in his doctoral thesis that the percentage change in the price of a stock in the French stock market due to speculation had the following properties: the random change in today's profit is independent of that of yesterday's profit and does not influence tomorrow's profit; each day's profit is small; the economic process underlying the change in profit is dynamic and complex, but its nature does not change from day to day so the statistics of profit fluctuations do not change; the day to day profit is additive. Given these assumptions Bachelier hypothesized that the movement in stock profit could be described by the law of error, that is, by the Normal distribution and the dynamics are given by the drunkard's walk.

The response of buyers and sellers of stocks varies from time to time in such a way that the market seems to be responding to a random force and behaves in a manner not unlike a speck of dust floating on a fluid being buffeted about by the many molecules of the fluid. The latter is the classical picture of the diffusion of milk in your morning coffee and is due to Einstein. In 1905, while still a patent office clerk, Albert Einstein published three of the most significant physics papers published in the twentieth century. One paper was on special relativity and revolutionized the ideas of space and time, while establishing the equivalence of mass and energy; another was on the molecular hypothesis providing the first self-consistent explanation of classical diffusion; and the third was on the quantum hypothesis explaining the photoelectric effect for which he subsequently won the Nobel Prize. It was unfortunate for Bachelier that although his analysis was the first to determine the dynamics necessary to generate a time-dependent Normal distribution, the giant of physics, Einstein, had independently done the analysis and suggested that the phenomenon was that observed by Robert Brown [29]:

> "It is possible that the movement to be discussed here are identical with the so-called "Brownian molecular motion"; however, the information available to me regarding the latter is so lacking in precision, that I can form no judgement in the matter."

However by his second paper on the subject the following year Einstein was convinced that Brownian motion was produced by the same physical mechanism that generates classical diffusion and included the term in the title of the paper.

The mathematical process now known as Brownian motion as noted above was named after the botanist Robert Brown [31] who in 1829 observed the erratic motion of pollen grains suspended in fluids, see Fig. (**2.4**). The actual effect was first observed half a century earlier by the Dutch physician Jan Ingen-Housz [32], who observed that finely powered charcoal floating on alcohol also executed a random motion. I introduce these historical nuggets to highlight the human side of science and point out that scientists are fond of memorializing one another by naming phenomena and/or theories after their discoverer/inventor. Unfortunately they occasionally get it wrong, so that it is perhaps more appropriate that Brownian motion be called Ingen-Houzian motion and Einstein diffusion should be called Bachelier diffusion. I know these name changes will not happen, but nevertheless it is important to be accurate and give credit where it is due.

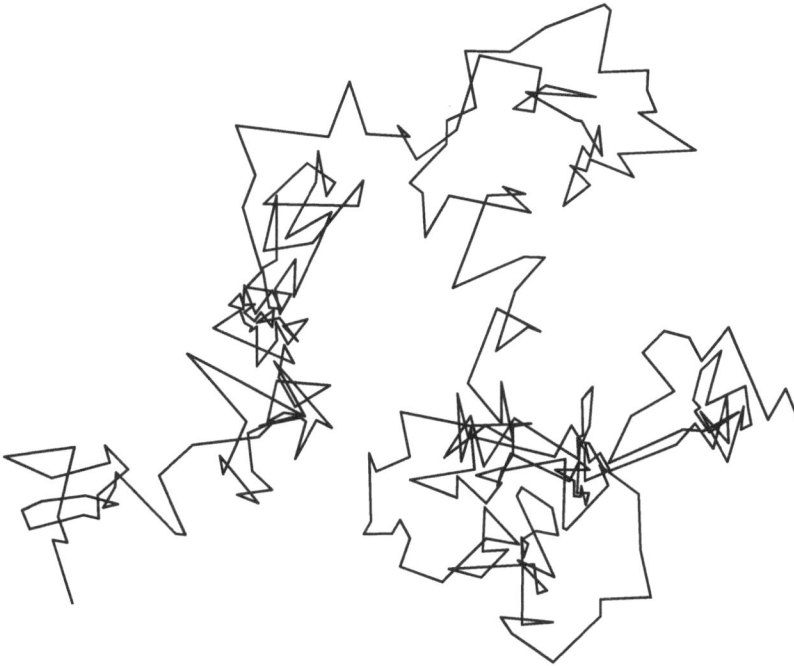

Fig. (2.4). The 'Brownian motion' of a heavy particle in a fluid of lighter particles. The points are the actual observations taken by Perrin [30] and the lines merely aid the eye is determining the ordering of the observations. The line segments do not indicate the location of the particle between observations. Perrin won the Nobel Prize for establishing the existence of molecules using this work. The erratic trajectory of the particle has been modelled using the drunkard's walk.

The law of error initially followed from a strictly mechanical view of the physical world, with the prediction of a dynamic equation being slightly but uncontrollably wrong. The error in the prediction is due to the fact that the simple models do not take into account the unknown and perhaps unknowable influence of the environment on the prediction. The bending of starlight and the buffeting of powdered charcoal provide physical examples of this system-environment interaction and in the social realm we have the fluctuations in the profit of a stock. The bell-shaped curve is a simple model that quantifies our uncertainty about the magnitude of the influence of the environment on the dynamics of the system. It is the combination of the Normal distribution being consistent with Newton's laws, with the ability to quantify the degree of error in any experiment that makes it so attractive. The fact that the error can be quantified suggests that there might be techniques that enable the experimenter to make the error smaller and smaller.

2.2. THE IMPORTANCE OF LINEARITY

In a slight breeze only small light things respond: ripples generated on the surface of a pond; leaves high up in a tree that catch the air currents; agitated water molecules that cool the skin as they escape into the sunlight; and the flag that opens and collapses with each puff of air. Larger things respond as the breeze stiffens; tree branches sway in unison; dust devils spring up alongside the road; kites pull and tug at their tethers. We learn at a very young age that the harder we push, the greater the response and bikes go faster when we peddle harder. All these experiences lend themselves to a linear view of the physical world and with age most people find that science reinforces these beliefs. For example, until the advent of satellite photographs the best predictor of today's weather was the weather over the last few days, with the influence of yesterday's weather being somewhat more significant than that of the day before. The weather was not alone in this regard; the future of almost every complex process was estimated by means of a linear extrapolation of its history into the future.

Fundamentally, the concept of linearity is necessary to unambiguously define what is meant by a variable in science. It is necessary because a variable can only be measured when the quantity of interest is either not influenced, or only weakly influenced, by the rest of the world. This situation can be achieved in the physical sciences, for example, within the typical range of room temperatures we can measure the length of a table. This implies that the coupling of the table's length to the physical properties of the room such as temperature is weak. However, it is not true that the two do not interact at all, since the temperature of the table is certainly the same as that of the room due to coupling between them. Newton knew that an object would eventually reach room temperature, a hot object cooling and a cold object warming, and he devised a formula based on temperature differences to determine the rate at which this temperature change occurs. That is how a modern medical examiner fixes the time of death (TOD) in a murder mystery; s/he takes the temperature of the kidney and uses the known rate at which the body's internal organs cool to room temperature to determine the TOD. The various rates at which objects cool down or heat up are determined by experiment and are recorded in tables. These are phenomenological constants and science has them in abundance.

2.2.1. Additive

In the linear additive world of Gauss, Adrian and Laplace, the average value dominates and that view was eventually accepted by a large fraction of the physical and human sciences. The linear view implies that phenomena are predictable; a small change in the present state of a process produces a relatively small change in its future behavior, with the output being proportional to the input. The conclusion is that physical phenomena can be controlled in a straight forward way; the world is stable and the appearance of instability is just that, an appearance, not a reality. But is that the world of stock market crashes, staggering unemployment and the failure of medical protocols; the world in which we live? To answer this question we first look at what the world is not and examine the cherished concept of linearity.

So what is meant by linearity and why is it important? The concept is important because a linear world has simple rules and yields simple results. In its most basic form, linearity determines that output is proportional to input, and response is proportional to stimulus. If I double the pull on my bow-string, my arrow flies twice as far. However, linearity arises in various guises in a number of distinct contexts and although they are in some sense equivalent, they do provide slightly different shades of meaning of the idea.

A child's swing is perhaps the most familiar example of a linear phenomena; if you don't push too hard. Energy is injected into the system by pushing and is extracted from the swing by friction at the points of support. The harder I push, the louder my son's laughter, as the swing arcs higher and further away. Gravity returns the arc of the swing back to my hands. The swing's motion is periodic, like that of a pendulum, as long as I push gently. The periodic motion is the same as that of a mechanical spring, which when stretched or compressed generates an internal mechanical force to restore it to its original state. The material world has been described by coupling together vast arrays of such springs to account for why boughs bend and eventually break, as well as, why musical instruments have such a broad spectrum of different sounds.

Let us try and understand more about this idea of coupling together linear elements to form linear networks. Consider a network consisting of a number of elements in which no one element is dominant. A central property of linearity is that the response to an external stimulation of each separate element is determined by the magnitude of the stimulation of that element. This is the property of proportionality. For example, if the constant of proportionality, also called the sensitivity coefficient, were 1/2 for each element, then the output of the network would be one-half the input. In a physical system the efficiency of the network is 50% and half the input is lost as heat and sound, and does not appear in the output. However this simple model of interacting elements has been used in other contexts with vastly greater implications.

The rationale for the United States sending troops into Vietnam that was heard repeatedly in the decade of the 1960's was that such action prevented that nation from adopting a communist form of government. The argument for the war was based on a simple physical metaphor. Each country in the region was represented by a domino standing on end in close proximity to its neighbor, forming a tightly configured chain, receding into the distance. Pushing the domino at the end causes it to fall (become communist) and strike its neighbor, which causes that domino to fall and thereby induces a cascade of falling dominos and government transformations. The domino theory was a simple linear model of a complex geopolitical situation with the virtue that it could be readily understood and communicated to a mass audience. Other theories of the time although certainly more accurate could only be understood by experts and did not have the same mass appeal. This is certainly an example of form over substance and the simplicity of the argument has nothing to do with its validity. It is merely an example of how the world's complexity is replaced and misunderstood using simple models.

2.2.2. Independent

Another property of linearity is that the total linear network responds to an action as the sum of the responses of the separate elements. This is identified as the independence of the separate elements. In this case each element could have a different sensitivity coefficient. Consider a student taking a number of courses where the applied force is the prescribed homework. The sensitivity coefficient could be the perceived level of difficulty the student associates with each course, with mathematics and chemistry having low values but English and poetry having high values (or perhaps the other way around). With this interpretation of the response coefficient, a student's anxiety level and/or self-esteem is determined by the additive nature of these responses. The effect of all the courses taken together can be overwhelming in some semesters. Note that the courses are assumed to be independent of one another; an assumption that is challenged in the next chapter.

One of the more fruitful ideas of the seventeenth century was that an algebraic function has a geometric representation. We used this recognition in the interpretation of the population growth curve in Chapter One. Geometrically a linear relation between two quantities implies that if a curve is constructed with the values of one variable on the vertical axis (say, food supply) and the values of the other variable on the horizontal axis (say, time), then a linear relation between the two variables may be given by a straight line as assumed by Malthus. Such graphs enable people to visualize the activity and communicate what is going on. A growing food supply is depicted as a line climbing upward with passing time, whereas a diminishing food supply is indicated by a line sloping downward. From year to year, given the variability of the weather and other factors, one would expect the line for crop abundance to rise one year and fall the next, possibly fluctuating up and down randomly over a long time interval. In such a case the food supply is said to be increasing if its average increases over time, neglecting the short time fluctuations.

In the discussion of population models there were curves representing the exponential growth of population given by Malthus. In that model, if the birth rate is larger than the death rate the line has an upward turn with the population growing faster than linear as time passes. If, however, the death rate exceeds the birth rate then the population takes a downward turn, eventually reaching extinction if things do not improve. Malthus used these results to argue that the policy of protecting the poor that inspired such laws as the *Poor Laws of England* were basically ineffective, because the source of poverty was a natural phenomenon. He believed that social policy could be based on such simple models; not so different from the domino theory of geopolitics proposed some two centuries later.

Linearity is essential for the assumptions that errors are additive and independent; necessary assumptions for the proof of the central limit theorem. Consequently the law of error necessitates the linear world view with its reductionist basis and the ambiguity produced by limitations in knowledge. This view was readily accepted by the nascent sciences of sociology, psychology and various branches of the life sciences. In this way the nineteenth century saw the birth of sociophysics, psychophysics and was the basis of a theory of medicine whose foundation was and to a large extent remains homeostasis. The concept of homeostasis, like the law of error adopts the view that there is a best value for the variable characterizing a physiological network. The heart rate estimated by the nurse, the blood pressure determined by the pressure cuff, and the breathing measured by the respirator are all average values; all presumed to give the best measure of the health of the physiologic network being probed.

This view of things provides an understanding of complexity that relies on a reductionist perspective. The strong form of reductionism states that understanding complex phenomena only requires an understanding of the microscopic (elementary) laws governing the underlying elements. This reasoning implies that once all the parts of a process are understood, they can 'added up' to understand the whole. In the physical sciences this procedure first took the form of the *Principle of Superposition,* which arises from the separation of a system into its fundamental linear components. In other words, like the law of error, the world consists of networks of linear additive processes, but unlike that law, these parts are inherently knowable. The linear world is predictable; it is Newton's mechanical clock.

2.2.3. Practical

The linear model is very useful in manufacturing, where the specification for the production of an object can be given to as high a tolerance as a machine can achieve. Of course no two production pieces are exactly the same and this variability must be taken into account. Suppose we are making widgets in which case the length is very important and the variability in length coming off the production line can be measured by the width of a bell-shaped curve. The goal of manufacturing is to reduce this width to as small a value as possible and thereby reduce the variability in the quality of the product. Implicit in this manufacturing model is the idea that variability is bad and must be suppressed. The success of the production line in western society reinforces the linear additive world view and the natural desire to control one's life conspires with it to make suppression of variability a human desideratum. But does this strategy for attaining uniformity capture a deep truth about the world, or does it merely impose an apparently desirable property on the world that is impossible to realize?

Nearly every college student believes in the linear world view to some extent, because they were introduced to it in their first large class, typically a survey course in one of the sciences. I recall being surprised to learn in my freshman chemistry class that the Professor would not give me an absolute grade based on what I learned, but instead he would dole out relative grades based on how much more or how much less I learned compared with the other students in the class. This was euphemistically called grading on a curve. The curve often left undiscussed is the infamous bell-shaped curve, with 68% of the students getting C's, another 27% of the students dividing B's and D's and the remaining 5% sharing A's and F's. But does this redistributing of grades accurately reflect the mastery of the material? Does the law of error have anything to do with the complex process of learning?

Consequently, just as the errors in a physical network can be tested to determine if real world data satisfy the Normal distribution, the same is true of data gathered from other disciplines. It should not matter if one is measuring the temperature in a room, a person's blood pressure or the heights of students on a university campus, the best representation of the data is the average and the variability in the data around the average is given by the width of the bell-shaped curve. This was the modern view of twentieth century science that eventually determined how the world ought to be understood.

2.3. THE NORMAL VIEW IS STRENGTHENED

Experimental data gathered throughout the nineteenth century supported the linear world view of Gauss, Adrian and Laplace. No one was more enamoured with the new perspective than the cousin of Charles Darwin (1809-1882), Sir Francis Galton (1822-1911) who compiled statistics on the properties of identical twins, the frequency of yawns, life spans, the sterility of heiresses, and the inheritance of physical and mental characteristics [33]. He introduced the bell-shaped distribution into the social sciences, as well as into the statistics of human measure and behavior. Sir Francis observed [34]:

> "I know of scarcely anything so apt to impress the imagination as the wonderful form of cosmic order expressed by the 'law of frequency of error'. The law would have been personified by the Greeks and deified if they had known of it. It reigns with serenity and in complete self-effacement amidst the wildest confusion. The larger the mob, and the greater the apparent anarchy, the more perfect is its sway. It is the supreme law of unreason. Whenever a large sample of chaotic elements are taken in hand and marshalled in the order of their magnitude, an unsuspected and most beautiful form of regularity proves to have been latent all along. The tops of the marshalled row form a flowing curve of invariable proportions, and each element, as it sorted into place, finds, as it were, a preordained niche, accurately adapted to fit it."

Galton's words reveal a sense of the awe and mystical wonder that some scientists had towards statistics, before the turn of the last century. Others were more pragmatic and treated the Normal distribution as a useful tool for calculation with no metaphysical significance. There was yet a third group that dismissed statistics out of hand, arguing that something that is wrong in each particular instance cannot be right in the aggregate. Regarding this last point, the Normal distribution has a regularity and stability characteristic of the data set as a whole, that is not the same as the variability observed in the individual measurements.

Manufacturers would find it difficult to plan an inventory if human body parts did not have a predictable distribution of sizes. The average shoe size certainly fits the greatest number of individuals in the population and those a size larger or smaller are a predictable fraction of this greatest value. When each small town had its own cobbler this was not particularly important and it was individual style and comfort that provided satisfaction. Each pair of shoes was customized and clothes were either made at home, or at the local tailor shop. One marvel of the big city was its ability to centralize shoe making and other things and be able to supply shoe stores with a large variety of machine-made shoes at low cost. The elegance of the cobbler's art withered, died and was replaced by practical replaceable footwear. This kind of replacement was symptomatic, or more accurately it was emblematic, of the industrial revolution, where the work of the individual was amplified by machine to enhance productivity. A hidden cost of making products of uniform quality and low price available to the general population was a compensating loss of individuality.

In order for these data to influence society it became important for the broader population to develop an intuition about the Normal distribution. It was clear that the underlying mathematics was not going to convince anyone without specialized training, so an alternative method of communication and explanation became necessary. One such method was the creation of a mechanical device that could act as an analogue computer to show the dynamical formation of the Normal distribution. Sir Francis constructed such a device as shown in Fig. (**2.5**) that actually mimicked a drunkard's walk, half a century before that particular mathematical model had been conceived.

The Galton board is the mechanical device sketched in Fig. (**2.5**) that can be used to generate a Normal distribution. On a large wood board a lattice of round pegs are constructed, with each peg in a given row being above the midpoint of the pegs of the row below. A glass pane covers the pegs so that the entire box can be stood upright and marbles released into the structure by means of the guiding funnel at the top and viewed through the glass. Each marble falls between consecutive levels and impinges on the center of the peg the next level down. Given the symmetrical arrangement, each marble has an equal probability of bouncing left or right as it strikes a peg. This is a mechanical realization of a drunk moving to the right or left with equal probability. At the termination of the peg region a number of channels have been constructed. A marble that has bounced out to a given horizontal position by the end of the peg region, falls freely within a given channel. Thus, most marbles fill the channels directly below the funnel opening, with increasingly fewer marbles filling the channels farther from this central position. The bell-shaped curve resulting from the dynamic free-fall of the marbles is evident.

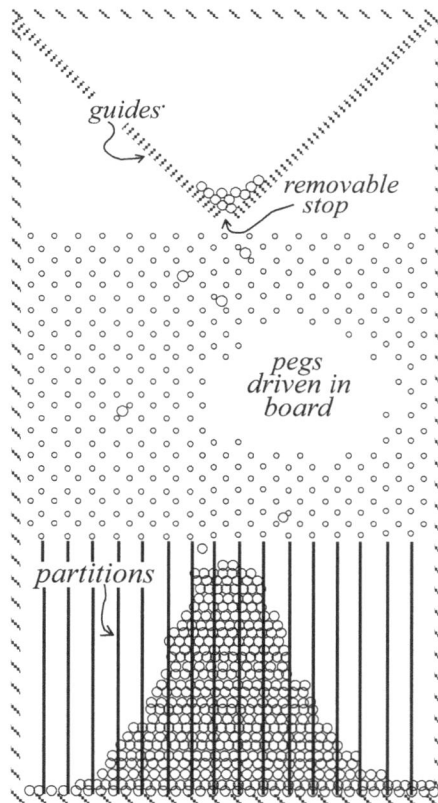

The Quincuni

Fig. (2.5). The Galton board, also known as The Quincunx or the bean machine provided a mechanical realization of the Normal distribution using a drunkard's walk. The sketch shown is a redrawing of a sketch made by Sir Francis Galton in 1898.

Each marble realizes a different path in travelling from the funnel opening to the channel where its dynamics come to an end. The marbles collected in the channels may also be interpreted as a collection of trajectories, each having the same initial condition, but travelling along a unique path to reach its destination. The static distribution of marbles displayed on the Galton board is a record of the sum over all the changes in the marble's path in time. Recall the discussion of light passing through the turbulent atmosphere with the resulting phenomenon of twinkling. The stacks of marbles at the bottom of the board could just as easily represent adding up all the trajectories of starlight as they erratically work their way through the turbulent atmosphere.

2.3.1. Sociophysics

During this same time period the observation regarding the variability of physical characteristics were assumed to carry over into the psychological and sociological properties of the individual as well. Adolphe Quetelet (1796-1874) helped develop the discipline of 'social physics', which patterned itself on celestial mechanics. Moreover, he used the law of error to explain deviations from the l'homme moyen, or 'average man' . He reasoned that human social variability was the result of error, much as Gauss and Adrian had argued for in the context of physical measurement [35]:

> "All things will occur in conformity with the mean results obtained for a society. If one seeks to establish, in some way, the basis of a social physics, it is he whom one should consider, without disturbing oneself with particular cases or anomalies, and without studying whether some given individual can undergo a greater or lesser development in one of his faculties."

The average man like the twinkling star is stable and therefore so is the society in which s/he exists. Social stability follows as long as individual variability is not too large and the dynamics of the social network controls variability through feedback, much as homeostasis was believed to do in medicine. This perspective was certainly put to good use by Quetelet's student Verhulst, as we saw earlier. Of course these assumptions completely overlooked the social unrest occurring at the time, such things as the American and French Revolutions, the War of 1812, and other such international quarrels.

Porter [36] points out that Quetelet's social physics was not a lasting contribution to science because it was in the final analysis unworkable. Quetelet's contribution was instead more cerebral in that he succeeded in convincing some illustrious successors of the advantage of considering the statistical regularities of society as a whole rather than concentrating on the concrete causes of the behavior of single individuals. Porter goes on to observe Quetelet's implication that the variability of the statistical approach was an immediate and general corollary of the universal truth that constant causes must yield constant effects.

Of course different disciplines gave this caricature of humanity different names: economics conceived of the 'rational man' and jurisprudence invented the 'reasonable man'. It is only now through the synthesizing of economics with psychology in neuroeconomics that people's irrationality finds its way into economic discussions, but the reasonable man is still alive and well in the legal system. Newmarch [37] sought to understand the condition of society through the laws of statistics:

> "...nothing less than the necessity under which all Governments are rapidly finding themselves placed, of understanding as clearly and fully as possible the composition of the social forces which, so far, Governments have been assumed to control, but which now, most men agree, really control Governments. The world has got rid of a good many intermediate agencies, all of them supposed originally to be masters, where in truth, they were even less than servants. The rain and the sun have long passed from under the administration of magicians and fortune-teller; religion has mostly reduced its pontiffs and priests into simple ministers with very circumscribed functions; commerce has cast aside legislative protection as a reed of the rottenest fibre; and now, men are gradually finding out that all attempts at making or administering laws which do not rest upon an accurate view of the social circumstances of the case, are neither more nor less than imposture in one of its most gigantic and perilous forms."

> "Crime is no longer be repressed by mere severity; Education is no longer within the control of the maxims which preceded printing, - Law is found to be a science perhaps the most difficult of any – Justice means more than

tricks and plausibilities of procedures; -Taxation, Commerce, Trade, Wages, Prices, Police, Competition, Possession of land, -every topic from the greatest to the least which the old legislators dealt with according to ...caprice...have all been found to have laws of their own, complete and irrefragable."

The linear statistical world view began to dominate the understanding of even the most complex phenomena in the nineteenth century. If the law of error could find application throughout the physical sciences then certainly the variability in the social sciences would be no less lawful. The application of these ideas to the field of medicine seemed most fertile, as we subsequently discuss.

2.3.2. Psychophysics

Francis Bacon (1561-1626) developed the concept that experimental science (empiricism) follow natural laws and that in so far as humans and society can be understood, in the same way as natural phenomena, they too must be similarly based, even if those laws remain a mystery. He reasoned that it is the existence of empirical laws and our knowledge of them that enables us to make predictions about the outcome of experiments and thereby obtain new knowledge. Consequently even before Newton (1643-1727) illuminated science, the world was seen as lawful and orderly, with regular patterns being the natural expression of nature's lawful character. Some believe that causality was the most important contribution of physics to the understanding of the world and that events, even those that are not yet subject to experiment, are lawful in their occurrence. Moreover, even when phenomena do not lend themselves to controlled experiment, data regarding their nature may still be obtained.

Efforts to understand complexity in human phenomena can often be traced back to the same scientists working to understand complexity in physical phenomena. In 1738, Daniel Bernoulli (1700-1782), the person who laid the ground work for our understanding of hydrodynamics and probability, also characterized the social well being of individuals by introducing the Utility Function. The mathematical function he choose to represent utility was the logarithm of the quantity of interest, because a change in utility or usefulness is then given by the fractional change in that quantity. Bernoulli reasoned that a change in some unspecified measure of value has different meaning to different people, depending on how much of the quantity they already possess. In this way a 10% change is more significant than a 5% change.

For example, if the quantity is a person's level of income, then the greater the level of income the less important is any particular change in income. Suppose I have a job with a yearly salary of $50,000 and you have one that pays $100,000 and that we both receive a $5,000 raise. Who is happier? We have both made the same additional amount of money, but it is an aspect of human nature, he reasoned, that we respond more strongly to the larger percentage change. You received a 5% raise and I a 10% raise, the empirical evidence is that I would be happier even though we have received the same absolute number of additional dollars. By the same token I would react more strongly on the negative side if we both received a similar pay cut. Experiments attempting to verify this aspect of human nature, occupied a significant number of scientists during the nineteenth and twentieth centuries, although it was taken as self-evident by Bernoulli, as indicated by his choosing the logarithm for the utility function.

The simple logarithmic form of the utility function cannot provide a complete characterization of an individual's response to external stimulation. However, it does seem to capture the essential scale-free nature of that response. Everything we see, smell, taste and otherwise experience is continually changing. The world is in a continual state of flux, but these changes are not experienced uniformly. Individual responses to the world's vagaries are not directly proportional to those changes. The nineteenth century physiologist, E. H. Weber, the brother-in- law of Gauss, studied the sensations of sound and touch and experimentally determined that subjects respond to the percentage change in stimulation and not to the absolute level. Shortly thereafter, the Physics Professor Gustav Fechner retired, and accepted a Chair in Philosophy and founded the school of experimental psychology called Psychophysics. He determined the domain of validity of Weber's findings and in so doing renamed it the Weber-Fechner law.

According to this law people respond to the relative change in stimulation, rather than to the absolute level of stimulation itself. This early work supported the intuition of Bernoulli, even though its basis was not social, but was instead psychological/biological. It appeared to those that studied these complex phenomena that scale-free functions are the more natural way to characterize psychological and social phenomena. But the effects are subtle and Fechner did not get it quite right. A century later Stevens [38] argued that the logarithm should be replaced by a power law; an idea first put forward by Plateau in 1872.

Fig. (**2.6**) displays the fit of the power-law response function to experimental data. The subjective loudness of a sound (brightness of a light or other stimulus) is depicted relative to a reference sound. A reference sound is given an arbitrary value and subsequent sound levels are estimated in terms of the reference sound. A second sound, perceived as half as loud, is given half the value of the reference number. Similarly a third sound perceived as twice as loud is given twice the value of the reference level. These numbers, for a variety of subjective *versus* the actual intensities of the sound, are depicted in Fig. (**2.6**). The straight line on this log-log graph, determined as the best fit to the data, clearly indicates a power law for the psychophysical function. Other excitations have been shown to satisfy similar power-law behavior and include the intensity of light, smell, taste, temperature, vibration, force on handgrip, vocal effort and electric shock among a myriad of others [39]. Each of these disparate complex phenomena has a positive power-law index.

It is interesting to note that on the doubly logarithmic graph paper of Fig. (**2.6**) the relation between the two variables being graphed is linear with the slope of the line segment being the constant of proportionality. Thus, in the logarithmic representation the relation between the variables is linear. Consequently whether the linear representation or the power-law representation is the more useful depends on the interpretation of the quantity being measured.

This particular application of the fundamental principles of physics to psychological measurement did not receive the same wide acceptance as socio-physics. A possible reason for this lack of enthusiasm might be that the experimenters actually obtained results that successfully tested theory and both the theory and results were incompatible with a linear additive world view. The scale-free nature of the power-law distribution, with its relative insensitivity to particular scales, was overlooked by the majority of the scientific community, with the occasional maverick in this discipline or that, marveling at its significance as we take up in Chapters Three and Four.

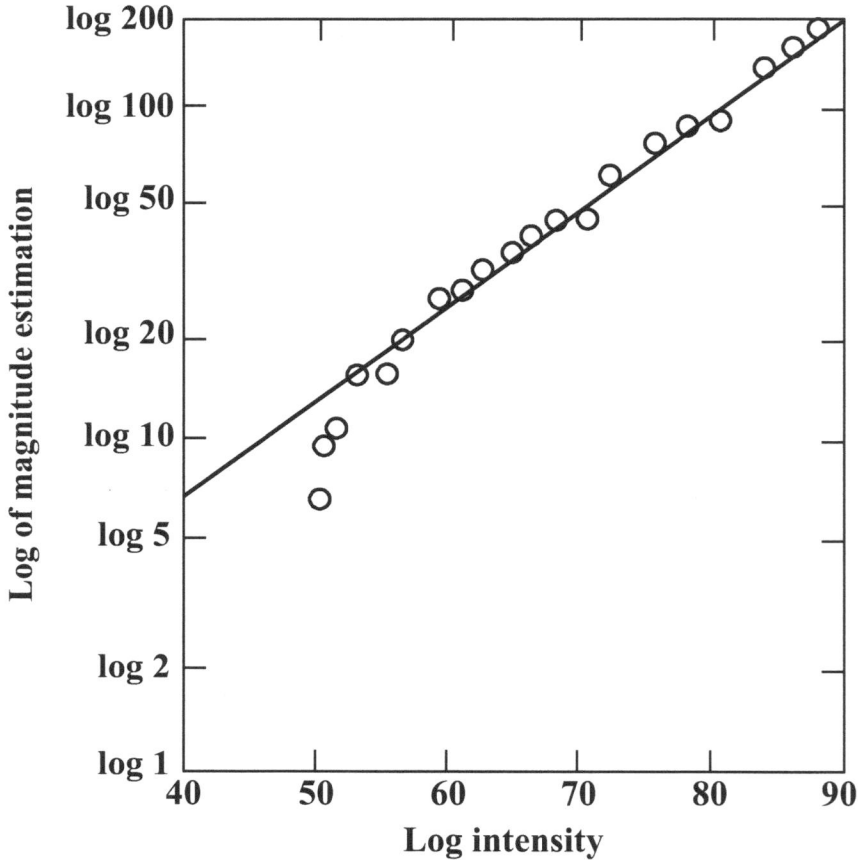

Fig. (2.6). The estimated *versus* the actual magnitude of a sound intensity is plotted on log-log graph paper. The solid line segment is the best fit to the data (open circles) [39].

2.3.3. Medicine

Florence Nightingale (1820-1910), see Fig. (**2.7**), was born at a time in history when young women of high society were expected to host parties, stand at their husband's side and graciously smile. However she had no brothers, one sister, and her father personally supervised their education, which included philosophy and mathematics. Florence's choosing to have a career was not favorably received by her father, but was reinforced by Dr. Elizabeth Blackwell, the first woman to qualify as a physician in the United States. In 1851 Florence's father finally agreed that she could be trained as a nurse. Two years latter Russia invaded Turkey initiating the Crimean War and British soldiers began coming down with cholera and malaria in amazingly large numbers. This "cholera epidemic" was so severe that of the 21,000 soldiers who died only 3,000 or approximately 14% were of the results of wounds received in battle.

Nightingale lead a group of thirty-eight nurses to Turkey where they discovered that hospitalized soldiers still wore their army uniforms that were "stiff with dirt and gore", and lacked decent food or blankets. Her efforts and eventual success in changing these appalling conditions and her fame as a reformer of hospital sanitation methods are unparalleled. What is not widely known was her innovative use of the new techniques of statistical analysis to quantify the evidence of preventable deaths during the Crimean War. She was able to directly relate the number of deaths to the lack of hygiene in the army hospitals. It is not an exaggeration to say she single handedly revolutionized the notion that social phenomena could be made objective through measurement and that such data could be mathematically analyzed to formulate new policy. It was not that academics and philosophers were not promoting these ideas, they were; it was that she applied statistics to the life and death situation of soldiers in war and conclusively demonstrated that lives could be saved through a proper interpretation of the statistical data. Moreover her family had the social connections to get the evidence to those in the hierarchy who could actually affect change.

Nurse Nightingale demonstrated that statistics provided a way to organize empirical information and learn about a complex phenomena to provide improvements in medical and surgical practices. She also developed a hospital form to collect, generate and achieve consistent data and statistics. Her contributions to statistical analysis lead to her being elected a Fellow of the Royal Statistical Society in 1858 and as well as an honorary member of the American Statistical Association in 1874, in part, for her efforts during the American Civil War. Karl Pearson (1857-1936) considered by many to have established the discipline of mathematical statistics, acknowledged Nightingale as a "prophetess" in the development of applied statistics.

2.4. LINEAR AND RANDOM

We mentioned that the desirability of knowing the future is part of being human. In Chapter One this concern over the future took the form of predictive models that motivated scientists to quantify phenomena and make very precise predictions of what the future might look like. The predictions are precise, but not always accurate; the accuracy has to do with the fidelity of the modeling, that is, how faithfully reality is represented by the model. When the discrepancy between prediction and measurement is systematic there is often an important mechanism that is overlooked in the model. Recall the deviation in the measured population growth of the United States and the saturation level predicted using the logistic equation. The discrepancy was accounted for by incorporating immigration into the growth scenario. So even when a model is wrong it can still be useful when its failure suggests a possible mechanism that was previously overlooked.

Fig. (2.7). In the 19th century Florence Nightingale used the nascent science of statistics to measure the effectiveness of hygiene in military hospitals in the Crimean War. She was able to conclusively show that the majority of the fatalities were due to unsanitary conditions and not to the soldiers' wounds.

2.4.1. Evolutionary Science

One strategy for improving an existing technology is to linearly increase the scale of its function; making it bigger or faster or stronger. The first recorded scientist to systematically embrace this strategy was the legendary Florentine Leonardo da Vinci (1452-1519). It is widely known that da Vinci was a painter, sculpture and scientist, but it is less well known that he made a significant fraction of his income designing armaments for Italian City States. His notebooks contain the first-ever designs of a tank, submarine and helicopter and he was without peer in projecting forward existing technology into previously unimaginable directions. His sketch of a giant cross bow in Fig. (**2.8**) indicated how to launch a bolt that could breach the wall of a castle; for a reference of scale size for the weapon note the man in the middle foreground. To appreciate the significance of this design you have to bear in mind that a dominant strategy for a lord to defeat an invading army was to shut off his castle and wait for the invading force to exhaust its provisions and ultimately leave. Any device that could open up the castle to the invaders was a game changer.

Fig. (2.8). A self-sketch of Leonardo da Vinci along with his sketch of a giant cross bow, note the man in the foreground [40].

This kind of evolutionary science, that leads to incremental advances in what is already known, has a ripple effect on technology. In Fig. (**2.8**) it is a linear increase in scale size; but what a phenomenal increase. The impact of such contributions are cumulative; often taking decades, if not centuries, for their realization in marketable technology. Most research in science and certainly the engineering applications of new science are incremental in precisely this linear predictive way through the development and refinement of simple models. Da Vinci's designs might have been implemented much earlier if his notebooks had not been encoded by him; he wrote backward, from right to left, without punctuation and in an idiosyncratic style. To further complicate the matter large parts of his writings along with the sketches were misplaced after his death for a couple of centuries.

The models developed in Chapter One incorporated the complexity of various phenomena into simple mathematical forms, some linear such as da Vinci's cross bow and others nonlinear such as Verhulst's population saturation. Perhaps the most unapologetic embodiment of the scientific view is attributed to Einstein on a plaque behind his statue outside the National Academy of Sciences in Washington, DC:

> "Everything should be made as simple as possible, but not simpler."

Of course the notion of simplicity is a relative one and what might be simple to Einstein's mind could be totally incomprehensible to the rest of us, but the principle embodied in his statement is clear. In the spirit of modeling adopted here, in the absence of evidence to the contrary that the simplest description of a phenomenon is linear. This is succinctly given by Occam's Razor:

> "Among otherwise equal explanations, the simplest is the best."

In the first section of the present chapter we examined how complexity, through uncertainty, blurs the predictions of the deterministic models presented in Chapter One. In the second section we discussed the reliance on the notion of linearity, even in the proof of the central limit theorem. Consequently, we combine the representation that enables us to construct a linear representation of a complex phenomenon and the random influence of the environment using Normal statistics. In such a model the output is nearly proportional to the input. The input to a system is under experimental control and is therefore known exactly. However there is uncertainty in the output of the system. It is the output that is measured and therefore cannot be known exactly. This type of modeling fills our lives and dominates how we think about the world.

This uncertainty is a somewhat formalized description of such tasks as asking your boss for a raise in salary. Under most circumstances there are elements of uncertainty in the response. On the other side of the social spectrum most children learn how to get a positive response from their parents. I remember telling my sons 'no' only to have them return a few minutes later with a more pleasant way of asking, or with an argument as to why my assessment was unreasonable. Of course many parents must endure the more unfortunate response of a tantrum. Such tantrums come in many forms; the yelling and screaming of a two year old in the supermarket followed by the falling to the floor; the dramatic altercation with your date at a party; and the inevitable clash over love among teens. My wife and I were thankfully spared the drama of childhood tantrums with our children. One son could argue better than either of us and the other could rely on charming his way out of any situation.

2.4.2. Inferences

In the linear world complexity manifests itself as the variability measured by the Normal distribution in the appropriate variable. The properties of the phenomenon being measured are then inferred from samples of the data population. This is statistical sampling and inference. When I say that a given fraction of a population is a minority, or is male, that describes a particular aspect of the population being represented. Males earning more money than females doing the same job, or minorities having teen-births with a greater fraction than their fraction of the general population, suggests that these phenomena and a myriad of others do not operate independently of the society in which they are observed and that societal factors play an important role in their development and persistence. These observations are then not merely statistical curiosities. Scientists seek to identify the possible mechanisms that can account for such effects and to suggest experiments and observations, when such experiments are not feasible, that may critically test the proposed explanations.

Have you ever noticed that the air is exceptionally clear after a rainfall; or heard that the number of mutations increases when living things are exposed to radiation; or experienced that respiratory ailments are more common in winter than summer. Causal relations are implied by associations such as these, but because the phenomena are typically not under our control there are always difficulties in distinguishing opinion from data. One such ambiguity is the effect changes in governmental policy have on the time evolution of complex social phenomena. In this subsection we discuss linear regression models that have been used more than any other technique for analysis of data from the human sciences for the past century.

The distribution of errors would not be of such importance to the discussion of how to transform data into information, if it had not had such a profound influence over the past two centuries on the development of concepts used to interpret phenomena in the natural and social sciences. Regression analysis can be traced back to Gauss' study of the calculus of errors. In this procedure the investigator starts by assuming a linear form for the trend in the data and models the deviations from the linear trend as a random process. One example of the application of this technique is indicated in Fig. (**2.9**). In this classic figure the rate at which the average energy is burned by mammals (basal metabolic rate) is graphed *versus* the mammal's average total body mass. It is incredible that a single linear curve on this log-log graph, the solid line segment in the figure, can connect what is happening within a mouse's physiology to that of an elephant, thereby giving a single linear relation that spans eight factors of ten in mass. The linear relation depicted in Fig. (**2.9**) implies that the trumpeting of the elephant and the squeaking of the mouse are not different in kind, but only in scale.

Of course not all species plotted in Fig. (**2.9**) fall exactly on the straight line, but that is the point. The trend or linear regression is given by the straight line and the data points that do not fall exactly on the straight line determine the fluctuations or randomness in the process. This particular relation between the logarithm of the average body mass and the logarithm of the average metabolic rate has fascinated biologists and ecologists for over two hundred years and is called an *allometry relation*. I return to allometry relations in subsequent chapters. For the time being take note that in order to make the regression relation linear the representation is in the logarithm of the data and not the raw data itself, which in this case would be the average basal metabolic rate and the average total body mass of an animal.

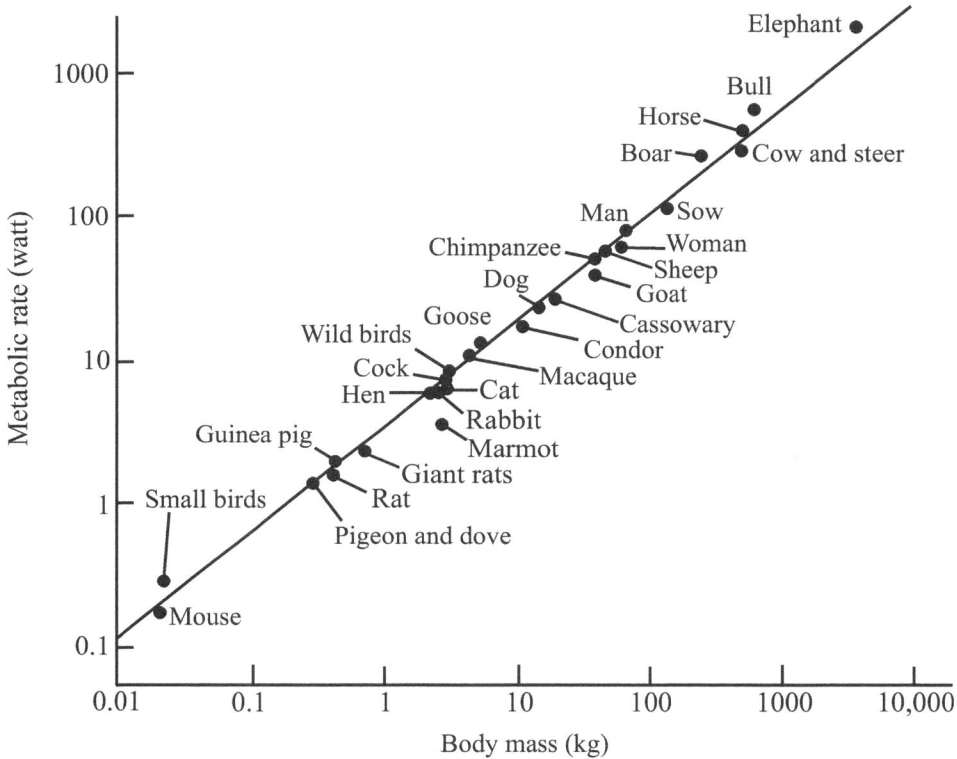

Fig. (2.9). The metabolic rate of mammals in Watts is plotted *versus* their mass in kilograms on log-log graph paper. [Adapted from [41] with permission].

The general linear regression procedure is to identify one variable as independent, this is the variable over which the experimenter has control, and the other as dependent, the variable that is being measured. A linear relation is assumed to exist between the two variables in terms of two empirical constants. An example might be the strength of the response to the amplitude of a stimulus, say the acceleration of a car to how hard the driver pushes down on the gas pedal. Early in his career Montroll studied the fluctuations in the acceleration of cars in a line where the cars were entering the Holland Tunnel going into New York City. He measured the fluctuations in acceleration in this car following configuration and determined that they had a Normal distribution. He and colleagues were able to construct a model showing that such platoons of traffic were invariably on the edge of stability [42].

A popular way to determine the properties of a data set is to use the drunkard's walk model introduced earlier. Suppose we assume the data points to give the length of the step in the walk. If each step is independent, like the falling marbles in the Quincunx, the variance would increase linearly in time. On doubly logarithmic graph paper the variance of the displacement *versus* time for classical diffusion appears as an upward going straight line with unit slope. If the slope is different from one then the underlying data points are not independent of one another. So let us examine some data.

2.4.3. Handicapping

I mentioned earlier that until the deployment of satellites the best prediction of weather was the previous days' weather. This is an extension of linear regression to time series that includes the influence of the past on the present. In horse racing this is called handicapping, where the history of the horse's breeding, her speed and pace along with her trainer and jockey are taken into account in picking a winner. If it is this difficult to pick a winning horse no wonder predicting the weather is so daunting. The systematic incorporation of history into the regression of a variable enables the past to forecast the future in the same spirit as predicting the saturation population from data using the logistic equation. The simplest way to implement linear regression, say in the case of weather, is to consider the high temperature for each of the past three days. Then average over all the previous days' high temperatures appropriately in order to forecast tomorrow's high temperature. There are a number of names for such procedures that incorporate the influence of a phenomenon's history into the forecast of its future. This is both the strength and weakness of the linear predictive measures, but for the moment we concentrate on the strength of the procedures.

Most of us muddle through life making decisions without complete information and in the face of uncertainty. In the modern world cause and effect are sought at every turn and therefore we make the same decisions in similar situations. If we lack reasons to change. Consistency follows from the expectation that if the cause remains unchanged then so too does the effect. Highway driving requires vigilance and a quick response to brake lights. However after the first snow fall a new set of responses must be adopted, as is well known by those who learned to drive in the northern states. Each year the failure to modify driving habits accounts for most of the cars in ditches during the first snowfall south of the Mason-Dixon line. The infrequency of ice and snow in the southern states insures that a large fraction of drivers native to the South will hit their brakes too hard or 'turn out of 'rather than 'into' a gracefully sideways gliding of their cars. In short, the rules have changed, but drivers do not remember what the new rules are, they make the same old decisions, which are exactly wrong; often with disastrous consequences.

People often use deviations from the input-output proportional response as an indicator that something is wrong. A casual greeting while passing a friend in the hall usually elicits a smile, nod or an equally innocuous response. However if the greeting goes unacknowledged, or the person weeps or snarls you might deduce that your friend is upset. This is where a decision regarding what constitutes a proportionate response enters the picture. A close friend might need emotional support, a sympathetic ear and some friendly advice. A colleague from work might actually react negatively to that level of response. Perhaps they require an expression of concern and acknowledgement that they are upset, but might consider advice intrusive. However you choose to handle the situation it is clear that it will be determined by how you have interacted with this person in the past and how you project that interaction into the present circumstance.

2.5. STATISTICS, SPECTRA AND SCALING

As I mentioned previously, in the late seventies a group of scientists, including myself, founded a non-profit research organization, called the La Jolla Institute. A major research focus of the Institute was ocean waves; how they are generated by the wind, how they propagate thousands of kilometers on the ocean surface and how they interact with ocean currents. The Institute had one office on Prospect Avenue in downtown La Jolla, overlooking the La Jolla Cove and another on the campus of Scripps Institution of Oceanography, where I occasionally taught. Each morning on my way to the Institute I would walk by the ocean, watch the waves roll into the shore and study the seals sunning themselves on the rocks below. At the time I considered this viewing as part of my research obligation.

The ocean surface provides an unmatched laboratory for observing the blending of the regular and the random into the ocean surface wave field. Waves are generated by the wind, not just the smooth regular flow of air, but the turbulent wind with its intermittent gusts. The raw wind drags the ocean surface from a smooth interface with the air into ripples determined by surface tension and increasingly higher amplitude water waves determined by gravity. The higher waves have longer wavelengths than the ripples and are organized over large regions of water. The windward side of the waves act as sails on which the wind pushes, not unlike the sails of a ship, except with water waves the size of the sail increases with time becoming more efficient at extracting energy from the wind as their height increases. The wind blows hard in some regions of the ocean and more softly in others, so the wave field being generated has a distribution of amplitudes determined in part by the spatial structure of the wind field.

The shoaling waves I could see breaking on the beach on my morning walks were clearly not uniform in height, but actually arrived in groups of low amplitude short waves followed by high amplitude long waves. These alternating sequences of waves are well known to surfers, who believe that the seventh wave is the highest, longest and will give the best ride. While the modulated character of the incoming waves is understood from the physics of oceanography, the number seven has less support and is part science and part folklore. The point is that even though ocean waves are extended over vast regions of the earth there is still local collective behavior that can be observed by those that care to look.

Consider the conditions of the central limit theorem, particularly the constraint that the errors are statistically independent of one another. Here the individual water waves are analogous to errors and they propagate with the smaller waves (errors) riding on top of the longer waves and interacting with them. This interaction makes the usually symmetric short wavelength waves rougher and more asymmetric near the crests of the longer wavelength waves. Thus, we would expect that the statistics of the water wave field is not Normal and that expectation is *almost* correct.

2.5.1. Oil Spill

The fact that water waves interact and generate localized structure lead us to the conclusion that the random water wave field is not Normal. However the interactions among these waves are weak so the deviation from Normalcy is typically small. Consequently some properties of the ocean surface are indistinguishable from a Normal wave field. The question is what properties are not Normal and how important are they? To answer such questions experiments had to be done. But it is very difficult to do experiments on the open ocean where almost nothing is under the control of the observer. One kind of uncontrolled experiment was to throw things on the ocean surface and record where they went. This tracking of objects would give an indication of the size, direction and variability of surface drift currents. In the middle sixties one quantity available in great abundance for such experiments were IBM cards. Recall that this was long before the takeover of the computer market by the personal computer and computer programs were punched onto such cards to be read wholesale into computers that occupied entire rooms, if not buildings.

A few boxes of these computer cards scattered from a bridge or boat would provide a network of tracers that could be tracked for long times and the trails used to determine the transport of matter by surface wave fields. An alternative to the computer cards were plastic cards, called drift-cards, that were used for the same purpose. Playing cards with Neptune was done quite often in the San Francisco Bay during the 1960s and 1970s. The hypothesis was that these cards would respond to the surface stresses produced by the wave field in the same way as an oil slick and they could be studied, without the attendant damage to the environment. The actual forces responsible for the dispersion of oil on the sea surface are stresses produced by water waves pulling on adjacent parts of the oil in different directions producing transport just as they did for the IBM cards.

Environmentalists and others were, and still are, very concerned with how rapidly oil spreads over the ocean surface from, say, the rupture of an oil tanker. This dispersal depends, of course, on the ocean wave dynamics that we have been discussing. It turns out that for a relatively low sea state the gross size and overall shape of a surface slick are well described by classical diffusion theory and hydrodynamics. The picture of the random walker introduced earlier to model the buffeting of milk particles by the lighter fluid particles in the cup of coffee is also useful for describing the movement of the heavier oil slick by random water waves. For water waves the random walker models the horizontal motion of the ocean surface due to the superposition of the passing nonlinear water waves. In this way the oil film coating the surface is pulled and stretched in every direction, thereby responding to stresses at the ocean surface. In fact the variance of the lateral size of an oil plume emanating from a point source is very well predicted by diffusion theory, that is, the transverse size of the oil slick increases linearly with the distance from the source of the oil spill as shown in Fig. (**2.10**). The variance of the oil spill has the same mathematical form as the diffusing milk in Einstein diffusion, with time in the latter replaced by space in the former. The straight line segment in the figure indicates how well the data points are fit by linear regression. Consequently for the purposes of predicting the extent of an oil spill the assumption that the surface wave field is Normal is adequate.

It is important to appreciate the significance of this last observation. We have used a model of the motion of 'heavy' molecules in a fluid of 'lighter' molecules, the kind of thing for which you would need a microscope to observe an individual event. However the macroscopic diffusion of milk, for example, is observable without a microscope because of the huge number of milk particles involved and their optical properties. The space and time scales are fractions of a millimeter and milliseconds for the individual events and centimeters and seconds for the macroscopic diffusion itself. The fact that the diffusion process is linear and additive makes everything work out. This mental picture scales up to the size of water waves that are meters in wavelength and seconds in duration forcing the diffusion of an oil slick on the scales of kilometers in size and days in duration. This kind of scaling with space and time suggests that the mental picture we have of the random walker transcends the process that gave it birth. When the phenomenon is linear, or nearly linear, the argument for Normalcy is independent of the phenomenon being considered. Our discussion of oil spills suggests that the ocean surface is on the border of Normalcy as long as the waves are not too large. For large amplitude waves we observe wave breaking, white caps and ocean spray indicating the increased importance of extreme events in the wave field and the loss of Normal statistics.

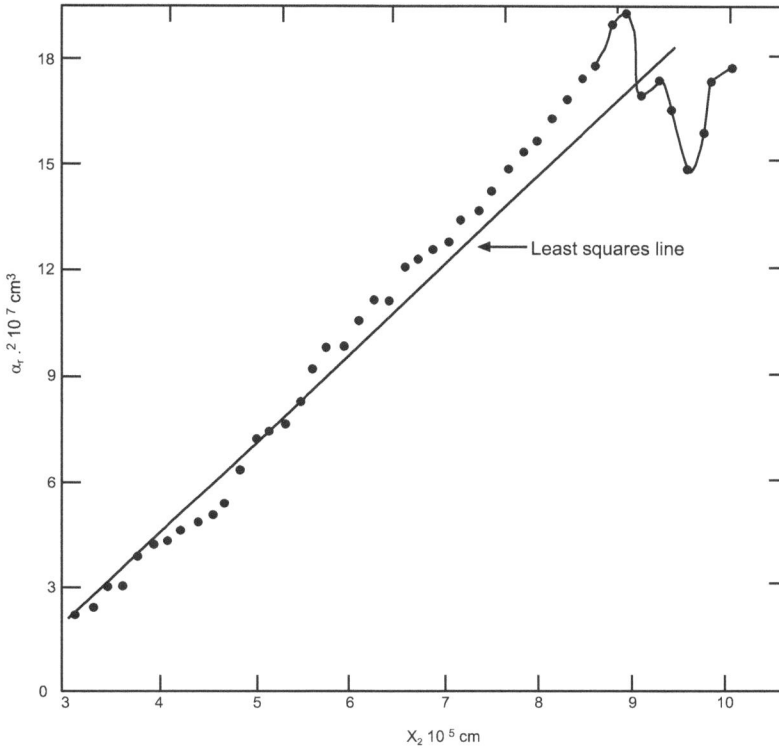

Fig. (2.10). The spreading of an oil slick on the ocean suface from the Chevron spill of March 11, 1970 in the Gulf of Mexico is measured as indicated by the dots. The variance of the lateral size of the spill in units of ten thousand kilometers is graphed as a function of the distance from the point source of the oil in units of one hundred kilometers. The straight line is the best linear regression to the data points and determines the diffusion coefficient for the oil spill. [Adpated from Murray [43]].

2.5.2. Normal Modeling

Despite its pejorative name the drunkard's walk was readily adopted by a number of investigators in a variety of disciplines as a formalism for incorporating uncertainty into the way processes and phenomena change in time. Consider following the motion of a woman in a red dress, as she maneuvers her way through the lunch hour crowd on a New York City street. From the top of a twenty story building she is just a red dot whose average motion could be described by a deterministic (Newtonian) trajectory. Her motion is not unlike that of one of those pollen motes seen under the microscope by Brown. The invisible causes of her apparently erratic movement to the right or left, as well as her speeding up and slowing down for no apparent reason, would all be modeled as a random walk in the vicinity of the drift trajectory. This separation of the average behavior from the instantaneous fluctuations appealed to modelers in a number of disciplines.

Neurophysiologists used the random walk model to help understand the discharge of neurons in the central nervous system. Sociologists found that the way rumors are transferred through social interactions could be modeled as the drunken transfer of information. Ecologists determined that the spatial spreading of new species in an ecosystem lent itself to random walking. Physiologists found that people undergo random walking even while they are quietly standing, because of the way the human body retains its balance.

The random walk model does not identify the mechanism that produces the fluctuations, but it does quantify such things as an average and variance, as well as, the size and frequency of the fluctuations. For the neurophysiologist the average is the rate of firing of the neurons and the variance is the level of variability in the number of neurons that fire in a given time interval. The expert then traces these empirical quantities back to the mechanisms that control or influence the dynamics of a neuronal network. In an analogous way the average rate at which a rumor is spread within a society is determined by the connectivity in the social network. The model is important because if it can faithfully capture the random variability of the process then certain conclusions can be drawn regarding its underlying properties. At a minimum the contributions to the variability must be linear, additive and independent just as in the central limit theorem.

In the sequel I outline why complex everyday phenomena violate all three of the basic requirements of Normalcy and more frequently than not, are nonlinear, multiplicative and interdependent. This violation is of concern because it implies that the random walk model cannot be used to understand most of the uncertainty in our lives.

2.6. REASSESSMENT

A historical investigative strategy adopted by scientists has been to restrict what they choose to understand. They make assumptions about the phenomenon being studied and predict outcomes within space and time intervals, where the assumptions are valid. For example a rubber band is elastic and springs back to its original shape if it is not stretched too far (the linear region). Thus, in the linear region I can confidently predict that the rubber band will oscillate periodically in supporting a small mass. However if pulled beyond the elastic limit it permanently deforms and eventually breaks (the nonlinear region). Consequently, unless I can determine the deformation properties of rubber I can only predict the dynamics in the linear regime. This simple example is meant to demonstrate that the control humans have over their physical world, as they do over their machines, is only a consequence of operating in the linear region. This controllability is part of what makes our mental pictures of the physical world so appealing. Even when uncertainty enters the picture, through random variability in response, control is maintained through knowledge of the probable behavior. The law of error guides our decisions because the future is not going to be too different from the past. The average response is considered to be an adequate description of the dynamics. The same is considered true in the social arena, but the evidence for that truth is more difficult to come by.

The philosopher John Stuart Mill (1806-1873) was one of the more vocal opponents to statistical reasoning. He did not think it possible that the human mind could construct a mathematics that would transform the uncertainty that prevails in the counting of social events into the certainty of scientific prediction. There are still many that today share Mill's skepticism and view statistics as an inferior way of knowing [44]:

> "It would indeed require strong evidence to persuade any rational person that by a system of operations upon numbers, our ignorance can be coined into science."

The simple world view of linear response and Normal statistics has been disrupted through the technological development of western society. Over time western societies have evolved, becoming more complex, less predictable and the levers of control have become more elusive. Society has become dependent on the transportation networks of planes, highways and railroads; the economic networks of global finance and stock markets; the social networks of businesses and religious organizations; and the physical networks of cell phones and the Internet [45]. All these interconnecting webs and more are neither linear nor Normal.

It is unrealistic to think of critiquing all the universally accepted linear models of crucial importance for the survival of society. Therefore I focus attention on one exemplar that is sufficiently familiar to most people, whose its importance is unquestioned and that is the medical concept of homeostasis. I choose this example because it is only recently that its efficacy as a foundational concept in medicine has been called into question. The notion of homeostasis was developed at the same time that sociophysics and psychophysics were being developed. Also at that time in chemistry the *Le Chatelier's Principle* was introduced, which states that when a system in dynamic equilibrium is acted on by an external stress, it will adjust in such a way as to relieve the stress and establish a new equilibrium.

The rate of relaxation is characteristic of the chemical composition being disturbed. The relaxation is part of a negative feedback process that ensures the chemical system does not stray too far from its equilibrium state. This is a stable network and at a time when many investigators were exploring new chemical reactions, it was important to be able to predict which reactions were stable and which were not, since the latter could lead to unexpected explosions.

Homeostasis is the medical application of *Le Chatelier's Principle* to physiological networks. The negative feedback in homeostasis is a linear dissipative process intended to ensure that the complex physiologic network remains at or near 'equilibrium', which is the defined state of health. Homeostasis is a significant part of the linear world view, with its simple feedback mechanisms producing the relaxation of a disruption back to the relatively quiet undisturbed state. However, in the real world, health is a state of great variability, not a quiescent state of equilibrium; death is the only equilibrium in life science. This dynamic range is necessary in order for the human body to rapidly adapt to short term changes in a complex environment.

The human body consists of a substantial number of interacting complex networks such as the cardiovascular, respiratory and motor control. The operation of each network is recorded by one or more time series, such as an electrocardiogram (ECG) that records the voltage amplitude associated with the beating heart. On a long time scale the time trace of the ECG looks regular as do the heartbeats. However, careful analysis reveals a great deal of variability in the inter-beat time interval. This high degree of variability is characteristic of physiologic time series and is quantified through the fluctuations in heart, breath and stride rates.

In Fig. (**2.11**) the data shown consists of the consecutive time intervals between heart beats of a young healthy adult male. If normal sinus rhythm did in fact represent the periodic beating of the heart it would be a flat line indicating that the time interval from one beat to the next is constant. It is evident from the figure that there are wild excursions of the heart rate and this is what scientists have found to be the usual situation. The present heart beat interval depends on those in the recent past and influences those in the near future and as you might have guessed, there are models of this correlation of heartbeats that are based on a modified drunkard's walk.

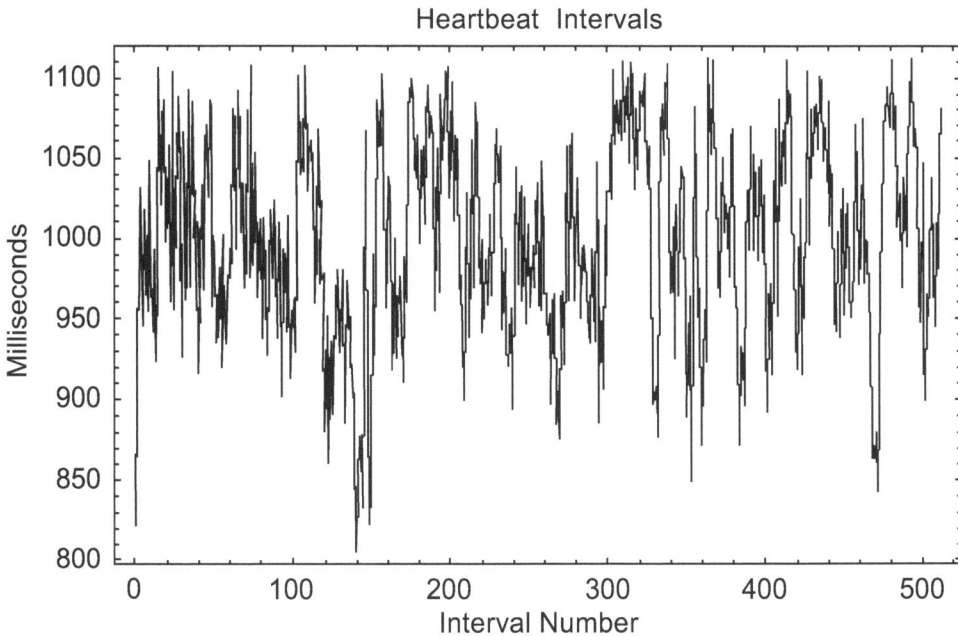

Fig. (2.11). The data indicates the change in the time interval from one heart beat to the next and is called heart rate variability. Lines are drawn to connect the dots in order to aid the eye in tracking the changes. The data is from a healthy young adult male. [Adapted from [46] with permission].

Unfortunately, the medical community long ago decided that there is no useful information in the fluctuations of the physiologic data. This proclivity to ignore variability was brilliantly summarize by Stephan Jay Gould [47]:

"... our culture encodes a strong bias either to neglect or ignore variation. We tend to focus instead on measures of central tendency, and as a result we make some terrible mistakes, often with considerable practical import."

So what do we use when the linear additive view breaks down and the average is no longer the important quantity? The answer given in the next chapter may surprise you.

In the linear additive world of Gauss and Adrian, the Utopian vision of Marx could be realized. Each person in that world would earn approximately the same amount of money, that being the average where the Normal distribution peaks. It is possible that some with exceptionally valuable talents would have a slightly higher salary and others that for one reason or another could not contribute fully would have a slightly lower salary. The average income would reflect the health and well being of such a society. Less than 2.5% of the resident population would be poor; the F region in the distribution of grades discussed earlier. In a symmetrical way less than 2.5% of the population would be rich; the A region in the distribution of grades. It is evident that a full 95% of the resident population would have income within two standard deviations of the average and in such a well-run society the standard deviation would be small. Consequently the poor would not starve and the rich would not live in obscene opulence. Who could possibly argue against such a vision of what ought to be true?

Of course this is not the world in which we live. But the linear additive model of the world developed in the nineteenth century suggests that such a world is not only possible, but is also desirable. In such a world one person cannot take unfair advantage of another because of their having an abnormally large intelligence quotient, or by virtue of being pathologically attractive, or through Faustian powers of persuasion, or for any other accident of birth such as their position in society. Individuals, as well as countries, worked passionately toward the realization of this and similar social visions resulting in the closing of debtor prisons, women suffrage, the ending of slavery, and many other positive changes, each one destroying a different societal inequity. Consequently, the linear additive world map was very useful in lowering the hills and raising the valleys in society's terrain; reducing barriers that had historically supported social control by one elite or another.

In this regard there was parallel development in the physical and human sciences, both relying on data and the exploitation of a linear additive world view. I use the term human sciences to include medicine and those parts of life and social sciences that pertain to human beings. This parallel development of physical and human sciences began to diverge in the twentieth century when the community of physical scientists began focusing more and more on complexity and the dominance of the nonlinear in interesting physical phenomena. It was not that the linear additive explanations no longer applied, it was that the frontiers of investigation into physical networks were advanced and the regions being explored were no longer those where the assumptions of the central limit theorem were valid. In examining complex physical phenomena Poincaré's cautionary warning regarding the inapplicability of the Normal distribution found fertile ground.

At the end of the nineteenth century the linear additive tapestry, into which social truths had been woven, began to slowly, but irreversibly unravel. It began with the collection and analysis of income data and the finding that the distribution of income was not fair, which is to say that it was not Normal. It had been known experientially that income and wealth were not fairly distributed in society, but the extent of that inequity was only quantified once the data had been gathered, analyzed and found to obey a law that was not Normal.

So how does the linear additive world differ from the nonlinear multiplicative one? We explore and explain this difference in the next chapter.

CHAPTER 3

Unfair: A Complex World View

Abstract: The linear additive world view, in which uncertainty is described by Normal statistics, is replaced by a nonlinear multiplicative world view in this chapter; the simple yielding to the complex. One consequence of the complex world view is that uncertainty is characterized by inverse power-law, rather than Normal, statistics. The implications of this complex representation of the world are immediate and profound. One inherent advantage is that the complex vantage point provides a single coherent view of disruptive mechanisms in complex phenomena; mechanisms ranging in physical science from earthquakes to floods; in social science from stock market crashes to the failure of power grids; in medical science from heart attacks to flash crashes in health care; and in biological science from the extinction of species to allometry relations. Extrema are more frequent in the complex world than they are in the simple world of Normalcy. The effects of extreme events are certainly unfair, and fortunately they do not occur every day. But when disruptive events do occur they introduce crossroads, and the selection of which road to take determines the subsequent course of events in a person's life. Consequently, understanding the source of extrema enables an individual to take back control from the hands of fate.

Keywords: Bursting, Complexity, Crashes, Extrema, Fractals, Hospitals, Intermittency, Nonlinear dynamics, Non-normal statistics, Power grids, Quakes, Tipping point, Unfairness.

One of our strongest urges as human beings is to know the future and through that knowledge control our destiny and the destiny of those in our charge. In the first two chapters I reviewed some of the historical evidences that the mental maps of the world we construct for ourselves consist primarily of elements that are linearly connected. Even when I introduced uncertainty into the description of events, as a way of including the influence of the broader world into their development, that uncertainty took the form of small additive random fluctuations. The world's ambiguity is represented by the distribution of the fluctuations in the outcomes of experiments and the variability of observations. The functional form of the distribution revealed certain general properties of the world's influence on simple predictions whether it is my estimate of the stopping distance, when tail gating at 60 *mph*, or how students react to a change in testing procedures.

In this chapter I examine how the neatly constructed linear world view has been challenged by the complexity of modern society. It is not the case that humans have changed how they construct their mental maps of the world. It is that the linear assumptions made in the past are no longer useful in guiding decisions made when social interactions are long range, multiple, and anonymous. I will indicate how the disintegration of simplicity disrupts our lives and leads to such things as the mismanagement of the health care system, particularly through the dominance of extrema when Normalcy no longer suppresses the outliers. Specifically I am concerned with the form in which the notions of fairness and equity, born in the social unrest and industrialization of the nineteenth and early twentieth century, survive in the data of the twenty-first century; or more accurately how they do not survive.

3.1. THE IMBALANCE

The Italian engineer/economist/sociologist, the Marquis Vilfredo Frederico Damoso Pareto (1848-1923), as the nineteenth century drew to a close, determined the empirical distribution of income within western society for the first time. Pareto had worked as an engineer in business until he was middle aged and with the death of his father and shortly thereafter his mother, he left the business world and after a brief hiatus took a faculty position in Political Science at the University of Lausanne, Switzerland. Being trained as an engineer he was convinced that the social sciences were amenable to the same logical-experimental reasoning as the Natural Sciences. Late in life he justified his perspective in *The Treatise on General Sociology* [48] with the following:

> "Driven by the desire to bring an indispensable complement to the studies
> of political economy and inspired by the example of the natural sciences, I
> determined to begin my Treatise, the sole purpose of which - I say sole and
> I insist upon the point - is to seek experimental reality, by the application to

the social sciences of the methods which have proved themselves in physics, in chemistry, in astronomy, in biology, and in other such sciences."

Fig. (3.1). Marquis Vilfredo Frederico Damaso Pareto (1848-1923), engineer, sociologist, economist and philosopher. He made several important contributions to social science, especially in the analysis of individual choice and in the study of income distribution, which he found to follow an inverse power law for high incomes.

Pareto, see Fig. (**3.1**), was among the first to have the modern vision of society as a network of reciprocal and mutually interdependent elements. Consequently, he viewed social change as a process of action and reaction to maintain the social order, analogous to the then emerging concept of homeostasis in medicine. He reasoned that all social systems are composed of individuals with a distribution of moral, intellectual and physical differences, together, with what he considered to be non-logical actions, resulting from psychic states of sentiment. He consequently determined that the distribution of wealth in western society was not completely random, that is not Normal, but followed a different, but consistent pattern he determined to be inverse power-law. He called the inverse power-law distribution "The Law of the Unequal Distribution of Results" [49] and referred to the inequality in his distribution more generally as a "predictable imbalance". The surprising result was that the distribution of income not only does not have a peak at the mean value, but it continues outward far beyond what one would reasonably expect from Normal statistics.

Pareto's imbalance is ultimately interpretable as being implicitly unfair and such inequity is invariably found in complex networks, as has been documented throughout the twentieth and into the twenty-first century. The marked difference between the bell-shaped curve of Gauss, Adrian and Laplace and the inverse power law of Pareto is schematically depicted in Fig. (**3.2**). The tail of the Pareto distribution extends far beyond the central region of the bell-shaped curve and indicates a lack of a fundamental scale, with which to characterize the process. The loss of a characteristic scale indicates a divergence of the variance in the underlying process. Recall that the argument for using the average value to characterize the erratic outcome of a series of 'identical' experiments was that the Normal distribution was narrow and therefore the average was empirically determined to be a good way to represent the data. The width of the Normal distribution is measured by the standard deviation. If the standard deviation becomes very large (diverges) the usual way of determining how well the average value represents the collection of data is no longer valid. In the distribution of income it would mean that the number of dollars between the richest and the poorest in society is much greater than the average income. This loss of a reliable measure presents a major problem for those attempting to understand the world through the use of experimental data and the application of Normal statistics.

One could argue that the distribution of income in the nineteenth century is a weak platform from which to launch a social theory that is at odds with Normalcy. There are many reasons why over a century ago such inequities might exist in the distribution of income and these imbalances can be corrected just as others have been. However, even with the graduated income taxes, the formation of unions, social security and other social/economic mechanisms, it appears that this particular imbalance has yet to be corrected. Some would even argue that the imbalance has increased over the last century, with more of the nation's wealth being concentrated in the hands of fewer people.

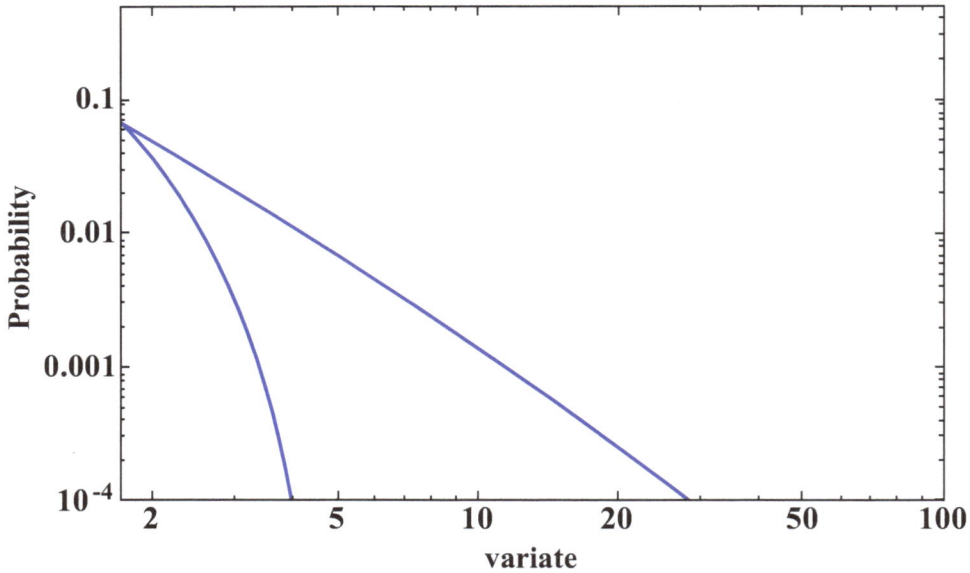

Fig. (3.2). The bell-shaped curve of Gauss, Adrian and Laplace is here compared with the inverse power law of Pareto on log-log graph paper where the Normal distribution is parabolic and the Pareto is a straight line. The long tail of the Pareto distribution is evident, indicating that large events are much more probable in the latter case than in the former and consequently much more important.

In Fig. (**3.3**) the distribution of income in the United States is indicated for 1980 and 1989 and Pareto's inverse power-law tail is clearly in evidence. It is true that the rich are different from you and me, particularly in how they use money. The economy of people in the upper 1% of the income level, as indicated in the figure by the Pareto distribution, is characterized mostly by investments. The economy of everybody else is determined by trade, where trade includes trading labor for wages. The existence of a middle class is also evident in the modern distribution and is here modeled by what is called a Gamma distribution, but how that part of the distribution of income is modeled is not relevant to the present discussion. What is of concern is the actual existence of a middle class, a group that did not exist during Pareto's lifetime and certainly was not present in his data. The inverse power-law tail that Pareto identified persists into the twenty-first century as does the middle class. The Pareto tail has not been legislated away, but the power-law index does change with the 1% level being pushed out to higher income across the decade of the eighties.

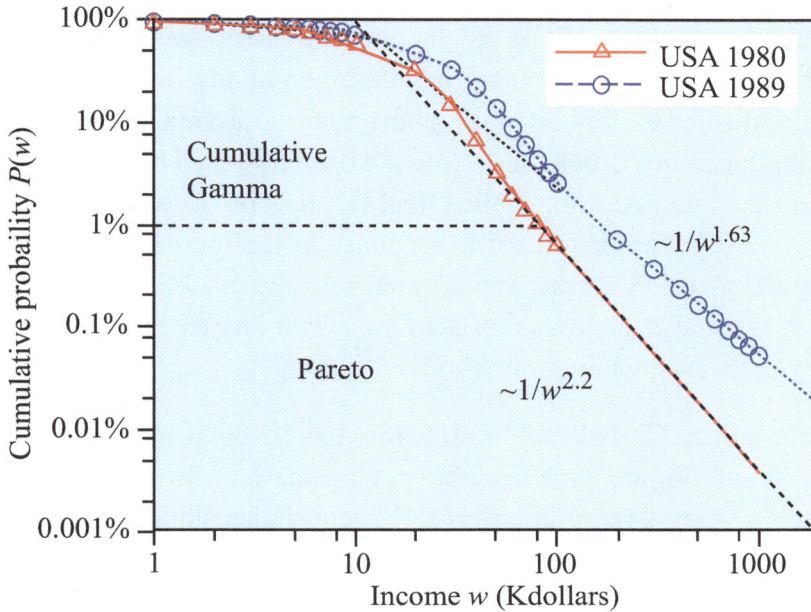

Fig. (3.3). Cumulative probability distributions for income levels in the United States for 1980 and 1989. The tails of both are described by the Pareto distributions with differing power-law indices. The income level ω is in thousands of dollars.

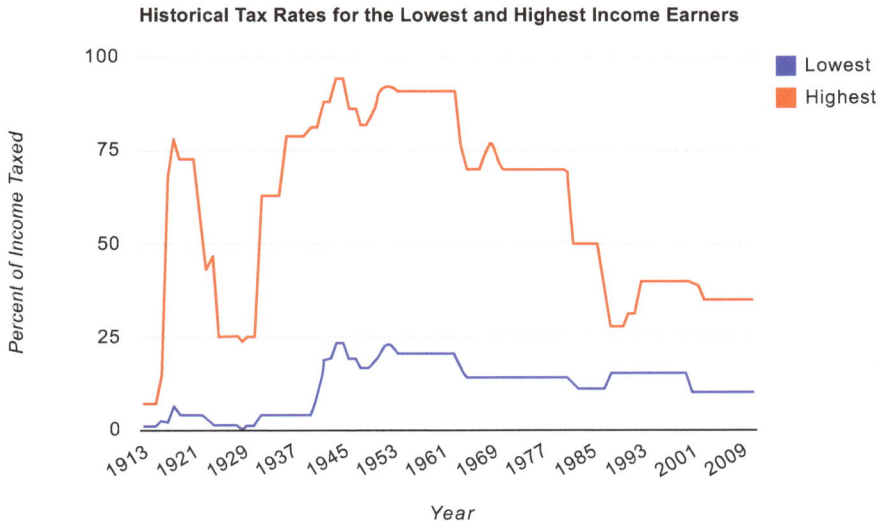

Fig. (3.4). The tax rates for the lowest and highest income earners in the United States over the twentieth century. [Adapted from [50] with permission].

The existence of the middle class may well be a consequence of the various social mechanisms put in place to lessen the imbalance in the income distribution and establish fairness. In Fig. (**3.4**) is shown one such social mechanism for establishing fairness and that is the graduated income tax. In Fig. (**3.4**) [50] the historical tax rates for the lowest and highest income earners in the United States for the twentieth century is depicted. Note the drop followed by the sudden rise in the maximum tax rate just prior to the Great Depression. The maximum tax rates of greater than 80% were instituted in response to the tragedy of the depression and were intended to reduce the social imbalance and establish fairness. What is remarkable is that these tax rates persisted for over a quarter of a century. So how well did they do in establishing fairness?

One measure of the imbalance in income level across society is the Gini Coefficient. Its definition is somewhat technical, but for our purposes it is sufficient to note that a value of zero for this coefficient indicates that everyone within the society has the same income and could correspond to one definition of social fairness. A value of one for this coefficient indicates that all the money has collapsed into the hands of the most wealthy, leaving the rest of society to starve. In Fig. (**3.5**) the Gini Coefficients for the United States post World War Two are indicated as calculated by David Roper and put online (http://www.roperld.com/ economics/IncomeDistribution.htm). A casual inspection shows that the index grew about 10% from 1945 to 1970, while the maximum tax rate was on the order of 80% and was relatively constant after 1970. The coefficient appears to be at most only weakly responsive to the changes in the highest tax rates indicated in Fig. (**3.4**) that dropped from approximately 70% in 1970 to around 30% in 2009. During this time of substantial change in the maximum tax rate the Gini Coefficient varied only a few percent and seemed to come down a little in the new millennium. The stability of the Gini Coefficient indicates that the rich and the poor have been moving together during the past 30 years without substantial change in the social imbalance contrary to what is often expressed in the popular media. Was that fair?

What happens to the individuals in the top 1% income level as their tax rate is changed so dramatically and the Gini Coefficient remains stable? In Fig. (**3.6**) the Pareto index for the power-law tail of the income distribution is recorded for the

100 years from 1910 to 2010 also calculated by David Roper. Over the first sixty years it appears that the index steadily increased, reducing the income range controlled by the individuals in the top 1% income level. The Great Depression appears as a downward blip in the power-law index, indicating a temporary constriction of the income span between the rich and the poor. It is interesting that this decrease in value occurs while the maximum tax rate is at an all-time low and is restored to its upward trend when the tax rate jumps back above 70%. After 1970, an apparently significant year for the United States, the Pareto index reverses direction and begins a steady decrease thereby increasing the income range controlled by individuals in the top 1% income level. If this index were interpreted in isolation the conclusion might be drawn that the income imbalance between the rich and poor was increasing from 1970 to 2010. However the Gini Coefficient in Fig. (**3.5**) indicates that this interpretation is not correct, because the imbalance remained nearly constant over this forty year period. Consequently if the range of income over which individuals in the top 1% income level have control increased during this period, as suggested by Fig. (**3.6**), then it follows from the constancy of the Gini Coefficient that the income level of the poor must increase during this period as well. The unfavorable conclusion is that contrary to population belief the rich and the poor both got richer in the past half century.

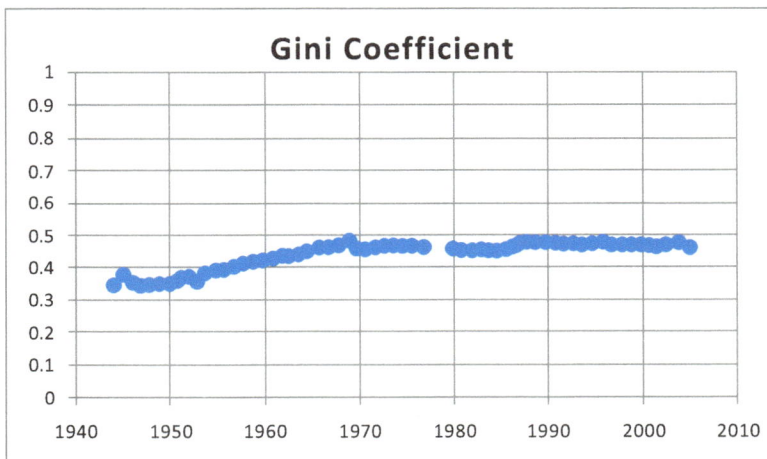

Fig. (3.5). The Gini Coefficient is calculated for each of the years folowing the Second World War using the income tax returns for the United States. A Gini index of zero has no imbalance and a value of one has a complete imbalance. [Adapted from [50] with permission].

A remaining question is whether the complete elimination of the Pareto tail in the income distribution is possible or even desirable? A number of mathematical models yield the result that the loss of imbalance in the distribution of income engenders social instability [51]. Of course a mathematical model is not proof, but then again neither is a passionately held belief. Both need to be tested using real-world data as done above.

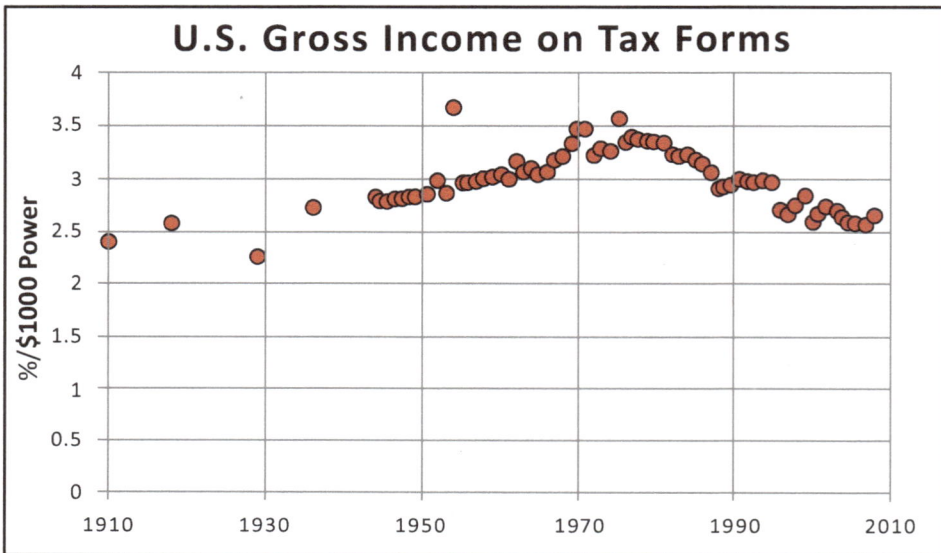

Fig. (3.6). The Pareto index for the income distribution in the United States is depicted for the twentieth century and beyond. [Adapted from [50] with permission].

3.1.1. Non-Normal Grades

Do you recall the universal grading curve we encountered in college? As a student I was never comfortable with being graded on a curve. I thought the curve unfairly relieved the professor of the responsibility of determining what the exact criteria to achieving a certain grade in a course ought to be. At that time I did not have data to test the assumption of Normalcy in learning, but that has changed. A few years ago I ran across a remarkable paper that included a data set that was used to test the Normality assumption in education. Fig. (3.7) records the data and analysis of Gupta *et al.* [52] on the achievement test scores of more than 65,000 students graduating high school and taking the university entrance examination of Universidade Estadual Paulista (UNESP) in the state of Sao Paulo, Brazil.

Of most significance here is the fact that the humanities data in the top panel of Fig. (**3.7**) supports the conjecture that the Normal distribution is appropriate for describing the distribution of grades in a large student population. The solid curves are the best fits of a Normal distribution to the data in the top panel and they appear to do quite well. The data was put into different grouping to determine if parental income level, or whether the students attending day or night school, significantly modified the qualitative form of the distribution. Although the location of the mean was shifted and the width changed for the different groupings, as one might expect, the distribution itself was clearly bell- shaped.

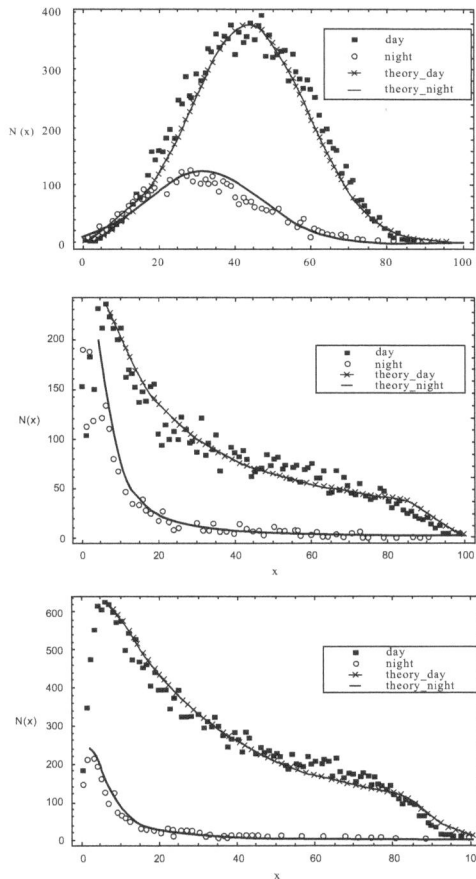

Fig. (3.7). The distribution of 65,000 student grades on the university entrance examination Universidade Estadual Paulista (UNESP) in state of Sao Paulo, Brazil: (a) humanities; (b) physical sciences; (c) biological sciences. [Adapted from [52] with permission].

The data recorded in the middle panel of Fig. (**3.7**), on the other hand, are from the physical sciences and graphed under the same groupings as that of the humanities. One thing is clear, the distribution of grades in the physical sciences is remarkably different from the Normal. The lower panel of Fig. (**3.7**) depicts the distribution under the same groupings as the first two, but for the biological sciences. The distribution for the biological sciences is more like those for the physical sciences than those for the humanities. In fact, the distribution of grades in the sciences is nothing like that in the humanities. The distributions in the sciences and humanities are so different that one would not associate unlabeled curves with the same general phenomenon, that is, the middle and lower panels would not be identified with learning. So is there a reasonable explanation as to why Normalcy applies to the humanities, but not to the sciences?

The distribution differences may well be a consequence of the structural distinction between the two learning categories. The humanities collect a disjoint group of disciplines including language, philosophy, sociology, economics, and a number of other relatively independent areas of study. Consequently the grades obtained in each of these separate disciplines are essentially independent of one another, thereby satisfying the conditions for the central limit theorem. In meeting the conditions articulated by Laplace the humanity grades take on the bell-shaped form.

Science, on the other hand, builds on previous knowledge. Elementary physics cannot be understood without algebra, and the more advanced physics cannot be understood without calculus, which also requires the understanding of algebra and more basically arithmetic. Similarly, understanding biology requires mastery of some chemistry and some physics. The different scientific disciplines form an interconnecting web, starting from the most basic and building upward, thereby strongly violating the independence assumption necessary to prove the central limit theorem. The empirical distribution for the sciences show extensions out into the tail region with no defined peak and consequently no characteristic value with which to summarize the data. The average values, so important in Normal processes, become almost irrelevant in complex networks such as learning. A better indicator of complex processes than the average is one that quantifies how rapidly the tail decreases in value. This decrease indicates how rapidly the influence of the tail on the central region of the distribution is suppressed and measures the strength of the imbalance in the distribution of accomplishments of the students. This interpretation is consistent with Pareto's observation that the distribution of abilities in people also has an inverse power-law distribution.

These data suggest that the Normal distribution does not describe the normal situation. The bell-shaped curve is imposed through education orthodoxy and by our preconceptions of how the world ought to be and is not indicative of the process by which students master information and knowledge, at least not those in the sciences. Thus, the measure of the pursuit and achievement of intellectual goals in science is not Normal. So how does this contribute to our changing view of the world?

3.2. NON-NORMAL STATISTICS

In Pareto's inverse power-law distribution of income a disproportionately small number of people have a disproportionately large fraction of the income. These are the individuals out in the tail of the distribution shown in Figs (3.2) and (3.3). The imbalance in the distribution of income was identified by Pareto to be a fundamental inequality and he concluded that society was not fair. The Pareto world view is very different from that of Gauss, Adrian and Laplace, as indicated by the two curves depicted in Fig. (3.2). It is also apparent from Fig. (3.7) that the same imbalance occurs in the sciences, where the truly gifted individuals have grades out in the tails of the distribution. So let us spend some time trying to understand the implications of these non-Normal statistics. We choose a few examples of complex phenomena whose statistics are not Normal, beginning with the distribution of earthquakes.

A word of caution before we begin however. The inverse power law arises in a wide variety of contexts and although they all have the imbalance noted by Pareto and others, the causes for the imbalance may be as varied as the phenomena in which they arise. There is one road to the Normal distribution and it is straight and narrow, if somewhat steep. However there are many roads to the Pareto distribution; some are wide and smooth, others are narrow and torturously convoluted. We explore a few of these roads with the hope of convincing you that their great variety is a consequence of the myriad of complex worlds in which we live.

3.2.1. Quake Sizes

Why can't scientists predict the size of an earthquake? For that matter why can't they predict when an earthquake of a given size will occur? The reason scientists cannot answer these and other similar questions has to do with the irregularity of such complex phenomenon. If the size of an earthquake had Normal statistics, the average would determine its magnitude and science would be able to predict, with some confidence, the likelihood of a given magnitude event occurring. Relatedly, once an earthquake of a given size occurs science would be able to predict how long we would have to wait before another of a comparable size occurs. The problem is that an earthquake is not the sum of a large number of relatively small independent geophysical events; once initiated a quake is a highly organized process that ties together the movement of vast amounts of land mass, called tectonic plates, with stress energy being intermittently released along fault lines. The fitful sliding of plates over one another, with their sticking and slipping, is experienced as intermittent quakes at the earth's surface.

The Richter scale measures the amplitude of the seismic waves generated at the epicenter of an earthquake, as recorded on a seismograph. The scale was invented in 1935 by Charles F. Richter, while working in collaboration with his colleague Beno Gutenberg at the California Institute of Technology. The number associated with the earthquake magnitude, say a 7 on the Richter scale, refers to a power of ten. Therefore two earthquakes that differ by one unit in magnitude, say from a 6 to a 7, are a factor of ten apart in size. It is the difference between crawling by a hidden police car on a downtown street at 10 mph *versus* roaring by him with your foot to the floor at 100 mph. On the other hand, the energy associated with a unit change on the Richter scale denotes a factor of approximately 32, or the difference between jumping off a one story building *versus* leaping from the roof of a thirty story building. A mouse might survive both falls, but a person would walk away from the first, and be splattered on the sidewalk in the second. A single unit on the appropriate scale can make a catastrophic difference.

The 8.9 magnitude earthquake that struck 78 miles off the east coast of Japan on Friday, March 11, 2011 was the largest in over 140 years. The average number of earthquakes world wide of a given size over the last century is shown in Fig. (**3.8**). Each axis on this graph is the logarithm of the indicated quantity, so an inverse power law would be indicated by a downward directed straight line segment, such as would result by connecting the points at the centers of the magnitude intervals. The slope of the resulting line segment is -1. It is clear from this graphing of one hundred years of data just how rare is a magnitude 8.9 earthquake. A 2.5 magnitude quake is a million times more frequent than is the 8.9 magnitude quake.

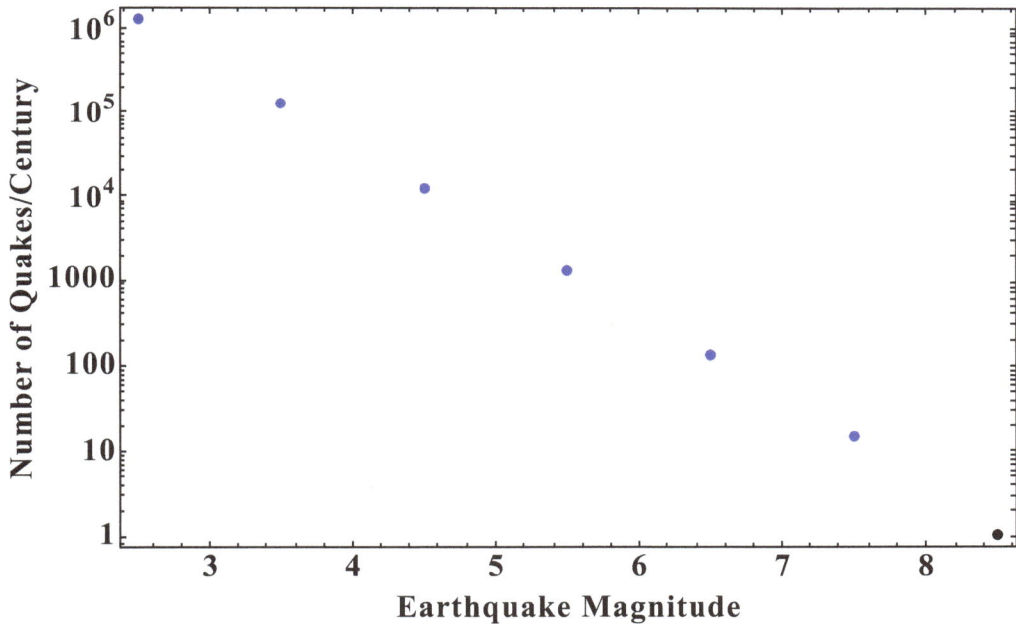

Fig. (3.8). Estimated number of earthquakes in the world of a given magnitude since 1900, from the *United States Geological Survey National Earthquake Information Center*. Connecting the data points would yield the Richter-Gutenberg law for the frequency of earthquakes of a given magnitude; this is an inverse power law with a slope of -1.

The Japan quake with all its devastation reduced the seismic risk of another earthquake of the same, or higher magnitude, occurring locally, but increased the seismic risk for such a large quake occurring along adjacent fault lines. It was once believed that earthquakes were independent events. However we now know that not only are aftershocks generated by large earthquakes, but major earthquakes along adjacent and even distant fault lines strongly influence one another.

3.2.2. Quake Times

How long one waits for an earthquake of a given size to occur depends on the probability of such an event occurring. In fact the waiting time depends on the inverse of the probability of the occurrence of such an event; very probable events occur frequently and we see them quite often, whereas improbable events are rare and may only appear to us in history books. Consequently, for a process described by Normal statistics the probability of an event of size far from the average is exceptionally small and therefore we would expect to wait a remarkably long time for such an extreme event to occur. In a Normal population extreme events are rare and data must be collected for a excessively long time before even one is seen. It is clear from Fig. (**3.2**) that the probability of an extreme event is much greater for a process with Pareto statistics than for one with Normal statistics. Therefore extreme events occur much more frequently in processes described by the inverse power law of Pareto than those described by Normal statistics. Using the data in Fig. (**3.8**) an estimate of the time between earthquakes of a given magnitude is very much shorter than would be predicted if these geophysical disasters had Normal statistics. Consequently if city planners used incorrect statistics of earthquakes, the building codes designed to withstand extreme quakes, thought to occur every 200 years, might instead occur multiple times within a single individual's life span.

Another approach to determining the length of time between quakes of a given size is empirical. Scientists have recorded when an earthquake occurs and measured the time interval between two quakes of the same size as shown in Fig. (**3.9**), starting with Omori in 1895 [53]. The Omori law is shown in Fig. (**3.9**) where the separate symbols correspond to different magnitude earthquakes and the data was originally collected from aftershocks generated by a major quake. The probability distribution for the time between quakes of the same size clearly has two regions; an initial domain from ten seconds to one year with a slope of -1, followed by a second domain from one to sixteen years with various slopes that depend on the magnitude of the earthquake. The region over which the -1 Omori law is valid increases with increasing quake magnitude.

Like the distribution of income, where those individuals at the high income end dominate the social situation, the extreme magnitude quakes dominate the geophysical situation. The damage done by the magnitude 9 earthquake in Japan dwarfs the effects of the much more frequent magnitude 7 quakes. The question of concern now is how the subsequent inverse power-law distribution in time are interpreted.

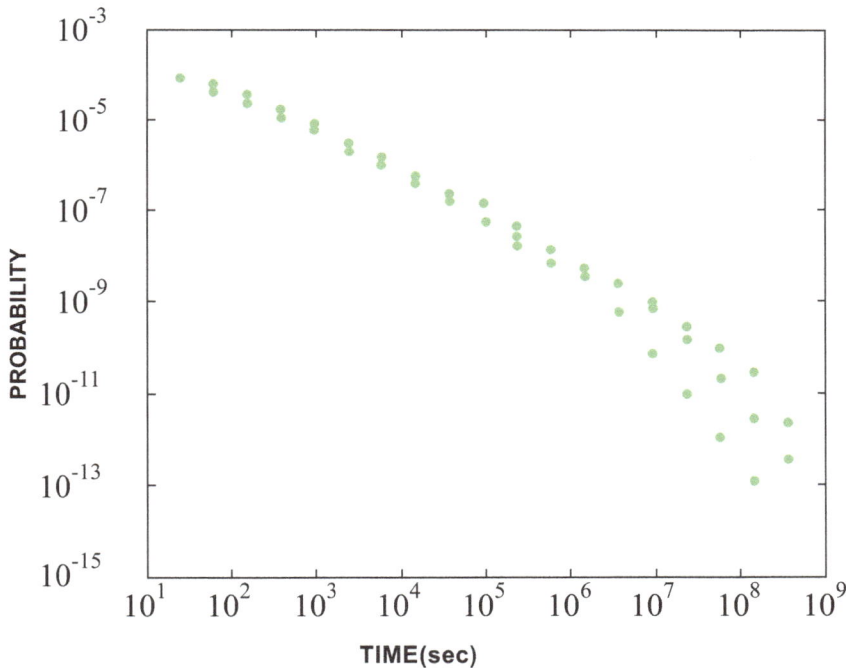

Fig. (3.9). The Omori law relates the probability of the time interval between two earthquakes of the same magnitude having a specified value. The region of validity of the Omori inverse power law increases with increasing quake magnitude. The triangles have magnitude 3; the squares magnitude 2 and the circles magnitude 1. [Adpated from [54]].

3.2.3. Power Grids

Aside from the safety factors required in building codes and perhaps some modifications in the design architecture in earthquake-prone regions of the country, there is little, if anything, that can be done about earthquakes, even when the statistics are right. On the other hand, power outages or blackouts are another thing altogether. A blackout has been defined as a power-network event that results in an involuntary interruption of power to customers and lasts longer than five minutes [55]. A blackout is not a localized event; the failure of a power grid can potentially affect the lives of millions of people and bring large regions of a country to its knees. In Fig. (**3.10**) the data from the North American Electrical Reliability Council (NERC) for 1984-2006 indicates the frequency of blackouts in the United States. In the figure the data is seen to track an inverse power-law distribution. Also shown is a well known theory of failure based on the assumption that the underlying process has Normal statistics. This latter theory is labeled the Weibull fit and deviates markedly from the data for extreme blackout sizes (a blackout in terms of the number of people affected or Megawatts (MWs) lost).

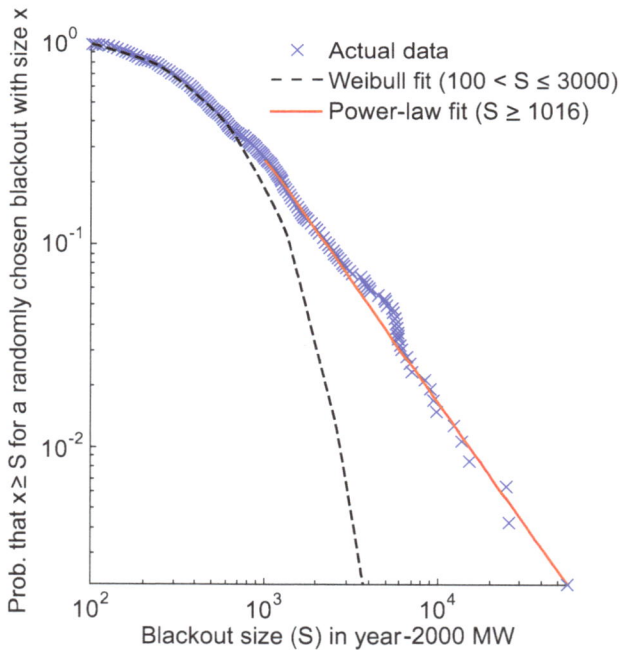

Fig. (3.10). The cumulative probability distributions of blackout sizes for events greater than S in megawatts (MW). The crosses are the raw data; the solid line segments show the inverse power-law fit to the data and the dashed curve is the fit to the Weibull distribution. [Reproduced from [55] with permission].

One would think that the frequency of large-scale blackouts ought to decrease over this time period, due to the advances in technology and the safety measures that have been put into place. Large blackouts typically involve complex cascading chains of events leading to failure that propagates throughout a power network by means of multiple processes. We emphasize that a particular black-out is explained as a causal chain of events after the fact, but that explanation cannot be constructed prior to the blackout. This is another way of saying that when a blackout will occur cannot be predicted. The uniqueness of individual blackouts means that policy makers cannot predict them in advance. In order to allocate resources and anticipate an event's effect policy makers must know the relationship between the size of an event and its probability of occurrence. The size of a blackout is determined by how many components of the grid fail and the probability of catastrophic failure is determined from the probability of failure of the separate components when the components are asymptotically independent.

Here again it is useful to consider the statistics of failure in a Normal population and contrast that with what is obtained in a complex network. This is what one finds in a simple manufacturing process such as the building of widgets. For 'widget' one can read any simple item that is mass produced such as light bulbs, fuses, batteries, *etc.* Empirically it is observed that for a large number of widgets there is a constant rate of failure per unit of time. This constant failure rate implies that the probability of a given fraction of the large number of widgets failing in a given time is exponential. Here that term exponential appears again, but this is not the exponential increase in population predicted by Malthus, this is an exponential decrease in the population of useful widgets. This is the population that survives to very long times. Although one cannot say whether this particular widget, or that particular widget, survives for a given time interval, it is possible to discuss fairly accurately what happens to a large number of widgets. It is useful for the factory owner to know the failure rate in order to anticipate what the manufacturing cost of the widgets will be over the next year and to redesign the production line to have an acceptable number of failures. A Normal population has the remarkable property that although the specifics of the process cannot be predicted or controlled, a single parameter such as the average failure rate can be modified to make the production process more efficient.

However, it is an inverse power law, not an exponential, that describes the size of blackouts as shown in Fig. (**3.10**) and the same distribution describes the survival probability of other complex networks. Consequently the control of failure does not rely on a single rate, as it would for a simple Normal process. No longer does the survival probability of an element or a network have the relatively rapid decline of an exponential, but instead failure has a heavy tail that extends far beyond the central region, making survival more likely, but increasingly more difficult to predict.

If earthquake statistics were Normal, the Omori law would be an exponential, in direct correspondence with the statistics of simple failure. However, neither the geophysics of plate tectonics, nor the control engineering of the power grid are simple enough to be described by Normal statistics. The individual parts of the separate processes are too interdependent for application of the 'simple' central limit theorem. So how can we use this insight?

3.3. FLASH CRASHES IN HEALTH CARE

One phenomenon that can be traced directly to the existence of networks and that could not have existed without them is the flash mob. Typically one or a few people text or email a substantially larger number of people to assemble at a given time, at a particular place, for a specific purpose, and then to rapidly disassemble. The first successful flash mob was generated by the author of the concept Bill Wasik in Manhattan in June of 2003 at Macy's department store. The Wikipedia records the event as follows:

> "More than 130 people converged upon the ninth floor rug department of the store, gathering around an expensive rug. Anyone approached by a sales assistant was advised to say that the gathers lived together in a warehouse on the outskirts of New York, they were shopping for a "lover rug", and they made all their purchases as a group."

Since that time flash mobs have popped in and out of existence at random locations to dance, sing and embarrass groups of bystanders. However most recently the purpose has become less benign resulting in 'mass shoplifting' where a group materialized in a store simultaneously, steals some items, and all head for the doors at the same time, making the likelihood of any particular person being arrested very small. This is but one example of the adaptation of social activity to communications networks. Another appears in a stock market instability called flash crashes, where the flash mob is a collection of stocks within a particular market, all these stocks rapidly lose a significant fraction of their value only to regain most of that value in a short time. The difference between a flash mob and a flash crash is that there is no one person or small group that is initiating the flash crash. Is there?

Clancy and West [56] all too briefly discussed the instability of the stock market that produced flash crashes and concluded that similar instabilities can and do exist in the United States healthcare system. This subsection is a modified and extended version of that discussion.

Fig. (3.11). The fall of the Dow Jones Industriabl Average on the afternoon of May 6, 2010 is shown as a 9.2% precipitous drop in the market, with an equally sharp recovery.

A leading indicator of stock prices, the DOW Jones Industrial Average (DJIA), plunged 999 points in less than 30 minutes, on May 6, 2010, as depicted in Fig. (**3.11**). Over $800 billion in value to investors was annihilated due to this unprecedented and unexpected drop in prices. This was immediately followed by the DJIA regaining $600 billion of its loss in just over 10 minutes, leaving investors reeling and Securities and Exchange Commissioners bewildered. Analysts continue to piece together what happened on that day, and a number of patterns are surfacing. As the Intranet has matured, vast electronic communication networks between multiple, interdependent domestic and international financial markets have emerged. Collectively, the structure of the network is tightly coupled and stock price information is instantaneously communicated to traders. Sophisticated computer models predict the direction of stock prices and automatically activate large blocks of buy and sell orders [56]. Because of extremely complex financial products called derivatives, stock price fluctuations amplify the sudden, unexpected changes in the economic environment.

There is an official explanation for the flash crash that begins with the unexpected selling of 4.1 billion dollars worth of contracts by a mutual fund complex that triggered an aggressive selling position .....and if you are interested the discussion is easily found on the Internet. But I do not believe the explanation.

Although the analysts providing the interpretation are saying what they believe to be true, they have fallen into a trap of their own making. The trap is that they think they understand how the market functions. The truth of the matter is that the stock market is a complex economic network that no one understands; least of all the generation, duration and magnitude of the flash crashes. Like the power-grid blackouts that no one can predict, there is always a flurry following the failure with an after-the-fact analysis. The facts of what happened in a network failure are put in chronological order and this ordering suggests causality to the analysts. However, correlation does not imply causality and because two events always occur in the same sequence does not imply that one caused the other. But it may suggest a relationship worth testing.

Such catastrophic events as flash crashes are inconsistent with the 'standard' model of efficient markets in which price variations are Normally distributed and statistically independent. The empirical evidence seems not to have swayed the finance orthodoxy as, observed by Mandelbrot [57] regarding an earlier similar event:

> "...by the conventional wisdom, August 1998 simply should never have happened...The standard theories... would estimate the odds of that final, August 31, collapse, at one in 20 million-an event that, if you traded daily for nearly 100,000 years, you would not expect to see even once. The odds of getting three such declines in the same month were even more minute:

about one in 500 billion...[An] index swing of more than 7 percent should come once every 300,000 years; in fact the twentieth century saw forty-eight such days."

An estimate of the probability of the May 6 event done similarly to Mandelbrot's estimate would be a mind-numbingly small number. The wild volatility in the stock market's flash crash on May 6th has remarkable parallels to the increasingly complex healthcare networks of today.

3.3.1. Hospitals

Both healthcare and financial systems contain connected webs of subnetworks that communicate through tightly coupled electronic networks. In healthcare networks these webs take the form of electronic medical records linked by an Intranet. As health networks have matured and become more complex, the flow of data has become more and more dependent on the network structure, that is, on the network connectivity. In complex networks that developed without overall design (such as the Internet), the distribution in the number of connections between elements often follow an inverse power-law distribution. Complex networks with inverse power-law distributions generate what Taleb called a Black Swan in his book of the same name [58]. A Black Swan is a single unexpected, unpredicted, cumulative event of enormous impact. This may, in part, have contributed to the flash crash the financial markets experienced on May 6, 2010. This could also happen in our health care system.

Black Swans probably happen more often than we care to admit. Consider the probability that a nursing units daily census suddenly jumps from its average occupancy of 60% to full use, 100% of bed capacity. If the percent occupancy on the unit is Normally distributed around a mean of say 60%, the probability of a jump to 100% occurring would be less than 0.3%. And that would be an accurate prediction if we only analyzed the midnight census, the usual hospital's report. This distribution generally follows a bell-shaped distribution as shown in Fig. (**3.12**). However this prediction would bump up against reality in short order.

The time interval between patients being admitted to nursing units is stochastic, which is a mixture of random and scheduled events as opposed to the purely random nature of the Normal distribution. For example, Fig. (**3.13**) presents the distribution of inter-arrival rate of admissions for a typical medical nursing unit. Note that Fig. (**3.12**) is a Normal distribution, while Fig. (**3.13**) is a distribution with a very long right tail; a realization of the Gauss- Pareto comparison in Fig. (**3.2**). Fig. (**3.13**) suggests there is a high rate of admissions with a very short inter-arrival interval (the tall columns on the left), interspersed with long intervals without an admission (the short columns on the far right). Consequently, there are "bursts" of admissions (during the day and evening shifts), followed by long quiescent periods of very few admissions (the night shift). The bursting of events is characteristic of time series with inverse power-law statistics.

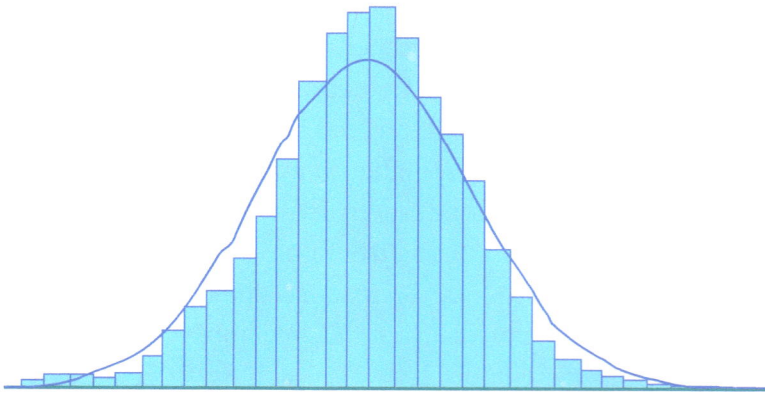

Fig. (3.12). Midnight census for typical medical nursing unit over a one year period. The peak is at 60% bed occupancy. [Adapted from [56] with permission].

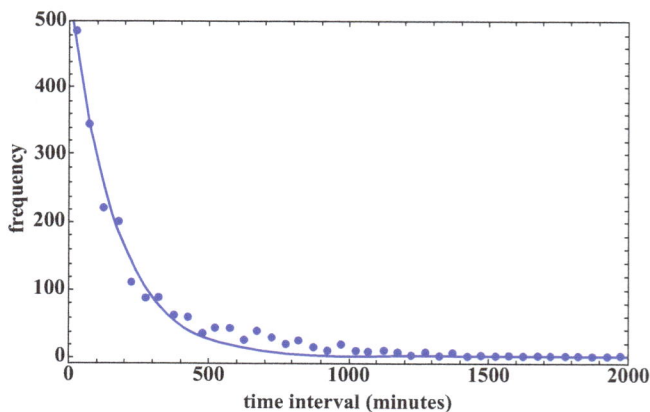

Fig. (3.13). Minutes between admissions onto a typical medical nursing unit histogram fit by an exponential distribution (solid curve). [Adapted from [56] with permission].

3.3.2. Bursts

Bursts of admissions can be visualized using time series plots. A time series consists of values of a dynamic variable measured and recorded at a given sequence of times; every time series is a discrete set of data points ordered in time. Fig. (**3.14**) presents the inter-arrival rate of admissions to the nursing unit depicted in Figs. (**3.12**) and (**3.13**) over a two week period and the line segments connect the data points to aid the eye in ordering the time intervals. Fig. (**3.14**) shows how the census fluctuates with admission rate changes over the same time period. Bursts in admission are represented in Fig. (**3.14**) as time intervals where a cluster of inter-arrival rates are near zero minutes. One of the bursts noted in this figure is from admission 12–16 (on the horizontal axis). Each data point is rounded off to the nearest minute.

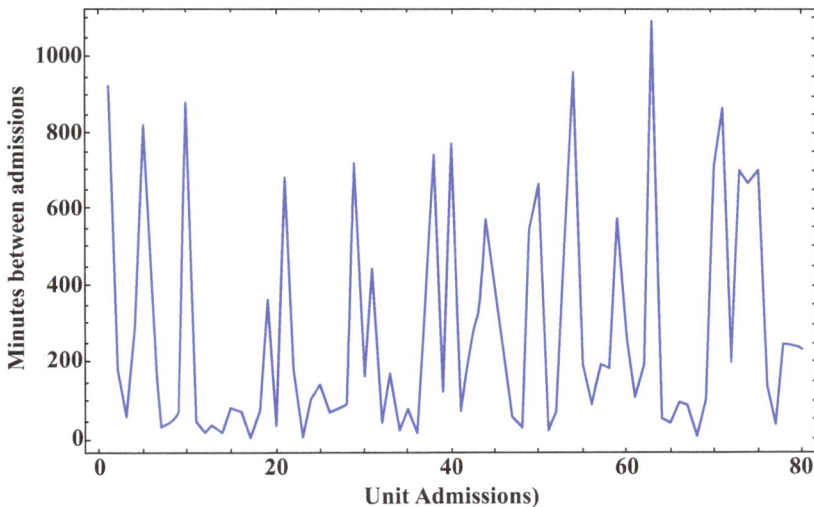

Fig. (3.14). Time series plot of a units census after every admission over a two week period. The lines are drawn to aid the eye in ordering data points in time. [Adapted from [56] with permission].

Notice in Fig. (**3.15**), that immediately after a burst, the number of potential occupants jumps to over 100% of bed capacity (25 beds). When the census spikes over 100% capacity, the excess indicates those patients admitted to the unit, but who are waiting for a bed to become available. The census peaks at 100%, or greater, far more frequently than that calculated from the midnight census would suggest (0.3%). These sharp spikes in census strongly correlate with nonlinear increases in nursing workload and represent the healthcare equivalent of a flash crash. In fact, Fig. (**3.13**) can be transformed to show the admission inter- arrival rate follows an inverse power-law distribution asymptotically, as shown in Fig. (**3.16**), similar to those seen in the financial markets just prior to their crash on May 6, 2010.

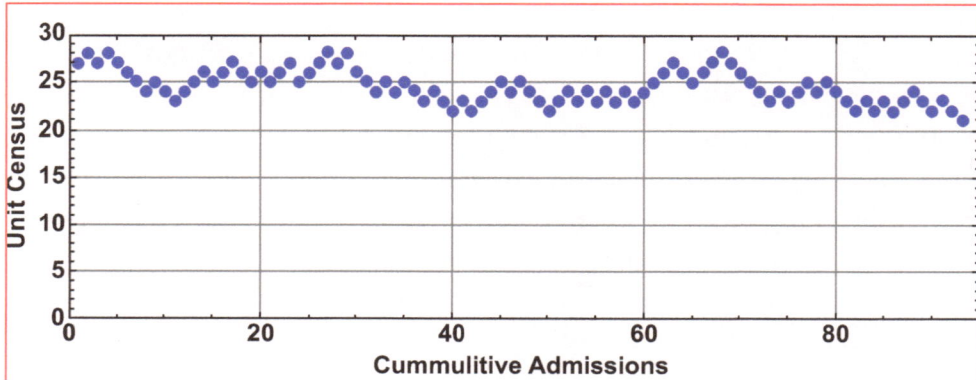

Fig. (3.15). Time series plot of a units census after every admission over a two week period. [Adapted from [56] with permission].

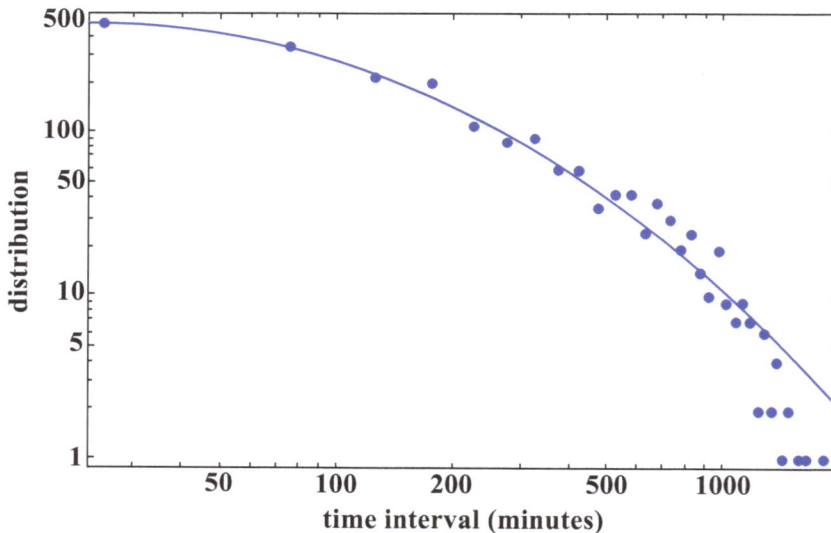

Fig. (3.16). The histogram on log-log graph paper for the time interval between admissions for one year of data. The solid curve is the best fit to the histogram data, called a hyperbolic distribution, and asymptotically becomes inverse power law. [Adapted from [56] with permission].

Like the stock market, the frequency of the occurrence of an earthquake of a given magnitude also has the form shown in Fig. (**3.16**). The challenge is to construct a building to withstand the 100-year earthquake, or to design a financial network that has safeguards to minimize the damage of the flash crash. Finally, in terms of the examples presented here, produce a healthcare network that can adapt to intermittent bursts of admissions.

Flash crashes indicate the increasingly interdependent nature of interactions, woven into the electronic and social networks of today's healthcare systems. They are the result of multiple factors converging simultaneously to unexpectedly and rapidly increasing nursing workload. Among these factors are a burst in admissions, electronic order transactions, and patient acuity. If these factors were random and independent, the probability of them all occurring simultaneously would be very small. Here the term simultaneous means that each factor occurs within the resolution time of the data; that would be one minute. Suppose the probability of occurrence of each of five factors was 1/2, then the probability of all five occurring together would be their product or $1/2^5 = 1/32$. This would correspond to 32 minutes between crashes. With a detailed under- standing of the process we could therefore estimate the time interval between such extreme events and be prepared for them. But don't put too much credence in the probability values from this last example The example is intended to indicate how such things are calculated when we do have credible probability estimates.

However, as shown in Figs. (**3.13**) and (**3.16**), bursts in census and workload spike quickly, but are never captured by the midnight census. Inverse power laws suggest a complex mix of underlying random and non-random mechanisms that drive network behavior. In the case of admission bursts, new evidence suggests that this mechanism might result from simply setting priorities, that is, be a consequence of putting the patients through triage.

It has been known for over half a century that the flow of entities (people, information, things) in human and other complex systems have a natural rhythm that is "bursty", which is to say intermittent in time. Whether it is the frequency of e-mails, automobile traffic, handwritten letters, or patients arriving at an emergency room, human activity usually occurs in bursts. And these bursts often occur in processes described statistically by an inverse power-law distribution. In *Bursts* [59], Barabási suggests that these bursts result from having to prioritize activities and wait in line. For example, in the case of a limited number of vacant beds on a nursing unit, those patients with the most critical needs are admitted first, while those with less critical needs are put into a queue. Subsequently, there are busts of activity (critical admissions), interspersed with long periods without activity (no admissions). What is most amazing about inverse power-law distributions is their frequency of occurrence, regardless of the application area. There appears to be an underlying principle at work, when prioritization is applied in complex human networks (such as in a hospital).

The existence of inverse power laws in complex networks has important implications for measuring performance, as, for example, in healthcare network behavior. Performance improvement methodologies that assume events are random, independent and Normally distributed, often underestimate the frequency of bursts, as well as, unexpected unpredicted events of enormous impact. In nursing networks these bursts might result in sudden spikes in workload or flash crashes. If flash crashes converge on multiple units simultaneously, the result may be a Black Swan, a catastrophic failure, such as a patient death from medical error.

In financial markets, computer models predicted that the probability of multiple markets crashing simultaneously was so small as to be virtually impossible; then it happened in the fall of 2008, when the United States suffered the worst financial crisis since 1929, only to be followed in May 2010 by an even more severe flash crash. However these highly improbable events were not entirely unexpected, scientists such as Gene Stanley at Boston University, who along with Rosario Manategna, invented *Econophysics* [60], had uncovered inverse power laws in financial data, and thereby anticipated the existence of flash crashes and Black Swans. However, since such 'laws' do not predict precisely when flash crashes occur, they were systematically ignored by the larger economics community and excluded from their dynamic models. The financial community did not develop any strategy to mitigate the effect of flash crashes and we are now reaping the consequences of that oversight.

3.4. VIEWS OF COMPLEXITY

The last half of the twentieth century witnessed a surge of interest in non-linear dynamics, first identified as a fundamental problem in understanding complex phenomena at the end of the nineteenth century by Poincaré. Like most other difficult problems, whether in science or elsewhere, a bumper sticker label emerged that captured what was not understood. In this case the label was the 'three-body problem'. I had not heard of the three-body problem as a graduate student taking the traditional courses in physics. It was only when I became a post-doctoral researcher, working for Elliott Montroll that I heard about it, even though its existence invalidated much of what I learned in graduate school about celestial mechanics. Until the 1970s the limitations of the methods for applying Newton's laws and solving the resulting equations were left largely undiscussed in the teaching of graduate physics. It was as if the limitation, engendered by complexity and embodied by nonlinear dynamics, was an embarrassment.

In addition to nonlinear dynamics, that entered the scientific discussion through physics, there were also scientists addressing complexity in the arena of ecology. The basic problem in ecology concerns how to understand and model the dynamics of multiple interacting populations. To explain the approach to understanding implemented in ecological science we discuss the dynamic balance between two populations, in which one is the food source for the other. This transfer of nutrition between populations introduces the idea of how networks influence one another and sets the stage for the discussion of information exchange. The latter exchange also requires a look into how to generalize statistical models made by Montroll and Weiss, into the domain of the inverse power law.

3.4.1. Prey-Predator System

Consider a relatively simple coupled system consisting of big fish and little fish, where the big ones eat the little ones. The growth equation for the number of little fish has a loss term linearly proportional to the number of big fish that eat them. A second equation describes the growth in the number of big fish with a gain term linearly proportional to the number of little fish that is their food source. If we start with a large number of little fish and a few big fish, the big fish eat the little fish, so the number of big fish grows and the number of little fish increases less rapidly than they would in isolation. When the loss in little fish due to being eaten by the big ones is sufficiently large the number of little fish begins to shrink. Eventually, with their loss of available food the previously growing pool of big fish begins to diminish. When the number of big fish is sufficiently small the number of little fish recovers and begins to increase again. Thus the number of little fish waxes and wanes and the number of big fish does as well, but slightly behind that of their food source. This is the much celebrated prey-predator model of which there are a number of historical mathematical renditions and empirical exemplars.

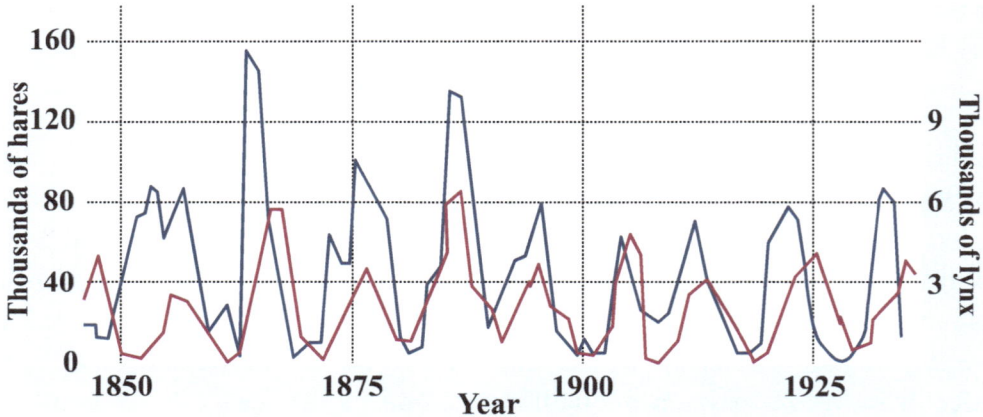

Fig. (3.17). The abundance of snowshoe hare and lynx across Canada obtained from the Hudson Bay Company records.

There are real world examples of prey-preditor systems in which the cycling of the two species is evident. An example of the prey-predator waxing and waning is depicted in Fig. (**3.17**) in which the numbers of snowshoe hare (prey) and lynx (predator) across Canada are depicted. The data was originally obtained from the Hudson Bay Company records, but I downloaded the figure from the Internet. The classical mathematical model that explains the oscillations in the two species case is due to Lotka and Volterra in which it is assumed that one species must eat the other to survive. As observed by Montroll and Badger [7] this model was motivated by a statistical analysis of D'Ancona of Adriatic fish catches over the period 1905-1923. The explanation of the observed variability in the fish catches fit the above fish scenario.

Of course complex networks need not grow deterministically as in the above scenario since their growth rates can vary sporadically and they can grow randomly in time. Another statistical phenomenon is spreading rumors within social groups that often rely on diffusion models, which implicitly contain random components that describe how individual pieces of information are transferred from person to person. Realistic decision making models incorporate elements of uncertainty to take into account incomplete information in the decision making process. Electroencephalograms show that the brain itself is not a deterministic engine of rationality, but fluctuates erratically in time, the degree of variability changing with the difficulty of the task being preformed. In fact the spectrum of brain response to difficult tasks is pink. Consequently it is interesting to question how one complex network responds to stimulation generated by a second complex network in the presence of random fluctuations.

The brain does a truly remarkable job of processing the information graphically displayed in Fig. (**3.17**). As we simultaneously scan the two population curves, we automatically line up the valleys and troughs and note a displacement, not a uniform displacement, but one that is erratic. It does appear that the lynx peak population precedes a decreasing hare population and that aside from some statistical fluctuations the prey-preditor model does fairly well in explaining why the oscillations occur. It would be useful to have a quantitative method to determine just how well the lynx population tracks the hare population, or conversely how sensitively the hare population responds to changes in the lynx population. Such a method does in fact exist and was designed to determine how responsive one data set is to changes in a second data set. This method is called cross-correlation. The measure is a cross-correlation coefficient when it has a single value. It is a cross-correlation function when its value changes with some parameter such as time. In the latter case the cross- correlation function measures the overlap between the stimulus and response signals as a function of time. This function has been used extensively to determine how complex networks respond to various stimuli in everything from human response to dose levels in drug trials to rehabilitation from traumatic injury to society's reaction to changing social policy. We return to its discussion in Chapter 5.

3.4.2. Generalizing Random Walks

Montroll was characteristically modest about acknowledging the depth of his contribution to particular areas of study. He was a genuinely interesting person, who had a keen sense of the past and regarded the history of a scientific problem at least as important as the contemporary understanding of that problem. His knowledge was broad and deep and in any given conversation he was the one laughing over the irony in the rediscovery of an important finding, or the misinterpretation of some result that was know a century earlier. From his casual conversations, or formal presentations, it was not possible to identify the fundamental nature of the contribution of his studies into the behavior of complex systems. Whether he was predicting the behavior of certain nonlinear systems, or revealing the statistical properties of networks in which the conditions necessary for Normal statistics are violated, it was the understanding that came through. His research had completely changed the fundamental understanding of such systems and yet his focus was on what we did not understand, rather than how his work had contributed to what we did understand.

In collaboration with George Weiss he changed how random walk theory could be used to understand complexity. They had examined the variety of phenomenon to which random walk theory had previously been applied and made a number of mathematically simple, but physically stunning observations. I take their observations out of the mathematical setting and put them into a sociological context, not to emphasize their simplicity, but more importantly to explain the wide ranging implications of those assumptions.

The first observation was that individuals in a society are not the same. Trivial you say? It was not trivial to the concept of the 'average man' in the nineteenth century, where the fluctuations were assumed to be small and the statistics to be Normal. In modeling the spreading of rumors sociologists had assumed that as far as the transfer of information is concerned all people are the same. This is what is euphemistically called a working hypothesis, or assumption in science and engineering; the intention being to go back and reexamine the problem without the assumption at some point in the future, after a better understanding of the phenomenon had been achieved. Unfortunately the diffusion model yielded too many interesting results and so sociologists did not look too closely at how sensitive the results were to the working hypothesis. Consequently Montroll and Weiss asked what is implied by not assuming that all people are basically the same but instead have a distribution of characteristics? One particularly important characteristic is with whom do people choose to share the rumor. In simple diffusion one tells only their nearest neighbors: What is the effect of telling people arbitrarily far away? The distribution of distances over which the rumor is transmitted in each transfer strongly changes the spatial statistics of the rumor. This assumption becomes particularly important in the modern world of tweeting, emails and YouTube.

A second consideration concerned time. It is a matter of experience that not all individuals transfer information at the same rate. My wife can take fifteen minutes to relate an incident that I would summarize in thirty seconds, or less. Of course, people much prefer talking to my wife, since she is by far the more interesting story teller. As to modeling the diffusion of rumors sociologists historically made the assumption that a rumor is transmitted between any two individuals in the same length of time. Here again the Montroll-Weiss investigation would ask what is implied by assuming a distribution of time intervals for transmitting information? How one models the distribution of times taken to transmit a rumor has a remarkable influence on the temporal statistics of the rumor. It is only in the case where the rumors have a characteristic time scale determining their transport that the statistics are Normal.

In short they had generalized the random walk for simple systems to a formalism that could handle spatial inhomogeneity, introduced by the differences in the characteristics of the elements in the process and the time delays, produced by the local dynamics of the elements. The resulting distribution of rumors need look nothing like a Normal distribution. But this is getting ahead of the story so let us return to the three-body problem and nonlinear dynamics.

3.4.3. Nonlinear Dynamics

The story actually begins with the King of Sweden, Oscar II, who in 1887 offered a prize of 2,500 crowns for anyone that could scientifically establish whether or not the solar system is stable. The King wanted to know if the planets would remain in their orbits forever, or whether the moon would tear itself from the earth's orbit, and the earth would crash into the sun. The winner of this contest was Henri Poincaré, who was not able to answer the question. However, he showed that only the two-body problem had periodic solutions in general, that is, in a universe consisting of only a single planet and a sun the planetary orbit would periodically return to the same point in space relative to the sun. This result is found in most first year physics texts. The remarkable and more important result, not found in those texts, concerns the motion of three bodies. Poincaré was able to show that if a third body is added to his hypothetical universe, and if that body was much lighter than the other two, then its orbit would not be periodic. This latter orbit has a very complex structure such as that shown in Fig. (**3.18**) and cannot be described by simple functions. One hundred years after Poincaré published his results it was determined that this orbit's structure is fractal. To Poincaré the complexity of the three-body problem indicated that mathematics could not answer the King's question. The understanding of the true nature of the three-body problem birthed what has become known as chaos theory.

Chaos theory captured the imagination of scientists in the last quarter of the twentieth century in part because it explained the behavior of nonlinear dynamic systems that manifest counter intuitive behavior. The allure for many was the previously unexpected behavior that occurred in the simplest nonlinear systems whose dynamics forced the reexamination of the notions of predictability and what it means to be deterministic. The unanticipated behavior is due to the fact that the chaos emanating from deterministic equations makes the outcome unpredictable. The process need not be complicated to be chaotic, that is, it can be small and compact with only one or two moving parts. It is neither the size nor the number of connections that is important for the emergence of chaos, just nonlinearity.

Laplace believed in strict determinism, which to his mind implied complete predictability given sufficient information. If the present state of a system is known with sufficient accuracy, we can predict its state at subsequent times with absolute certainty. For Laplace uncertainty was a consequence of imprecise information, so that probability theory was entailed by incomplete and imperfect observations, such as would lead to the law of error and the central limit theorem. Poincaré, on the other hand, saw an intrinsic inability on the part of nonlinear systems to make long-time predictions, due to a sensitive dependence of the final state of a system on the initial state of that system. This sensitivity emerged even though the system is deterministic [25]:

Fig. (3.18). The trail left by the trajectory of a very small planet is depicted under the influence of mutual gravitational forces from two suns after a very long time.

"A very small cause which escapes our notice determines a considerable effect that we cannot fail to see, and then we say that the effect is due to chance. If we knew exactly the laws of nature and situation of the universe at the initial moment, we could predict exactly the situation of that same universe at a succeeding moment. But even if it were the case that the natural laws had no longer any secret for us, we could still only know the initial situation approximately. If that enables us to predict the succeeding situation with the same approximation, that is all we require, and we should say that the phenomenon had been predicted, that it is governed by laws. But it is not always so; it may happen that small differences in the initial conditions produce very great ones in the final phenomena. So small error in the former will produce an enormous error in the latter. Prediction becomes impossible, and we have the fortuitous phenomenon."

In the contemporary world of Poincaré and Pareto small changes can have large, if not catastrophic effects. The output of a network is not proportional to its input and the response is not proportional to the stimulus. In fact, simple but nonlinear rules applied to inputs lead to complicated outputs, so complicated that they are virtually unpredictable and we have the phenomenon of chaos as we mentioned in Chapter One. The circumstances attendant to the input must therefore be known to a high degree of accuracy to determine the output and very often even this degree of detail is insufficient for reliable predictions for any significant length of time.

It is very tempting to explain the technical reasons why the nonlinear dynamics explored by Poincaré lead to a new branch of mathematics and a very different way of understanding the dynamics of the physical world. Reasons that explain why all the smooth trajectories of traditional celestial mechanics are actually very rare when considering the full range of possible motions of interacting bodies. However I will take a deep breath and restrict my discussion to the single qualitative term - fractal.

3.4.4. Fractals

The word fractal was introduced into the scientist's lexicon in the last century by Benoit Mandelbrot (1924-2010) who I first met as a graduate student in the late 1960s. He was a friend of Montroll and would come intermittently to give lectures in the Physics Department. However his talks were remarkably different from those of any other speakers of the time. Mandelbrot's second book on the subject,

The Fractal Geometry of Nature [61], has become a classic with over 20,000 citations. A citation to a scientific article or book generally means that the author citing the work has found it to be useful in their own research. Of course there is also the situation where the author finds fault with the previous work and that is the reason for the citation. For whatever reason the number of papers having a given number of citations is an inverse power law as shown in Fig. (**3.19**). The average number of citations to all scientific articles published in a given year is between three and four and the average occurs beyond 96% of all published scientific papers. Consequently, no more than 4% of all scientists in the world have their papers cited four times or more in a given year. Mandelbrot's book is truly an exception with citations approximately 300 times more frequent than that of the average paper after more than two decades. This level of citation indicates that his fractal notion electrified a generation of scientists.

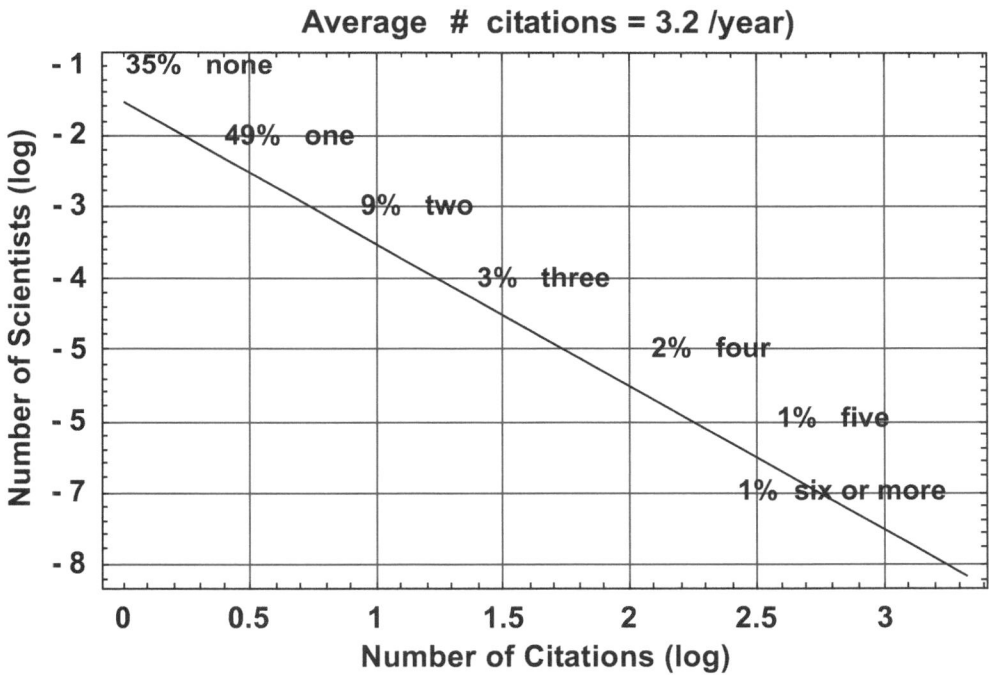

Fig. (**3.19**). The number of scientists having a given number of citations on a log-log scale. Note that the average number of citations is 3.2 per year so that on a logarithm scale this is 1.16, which occurs after 96% of all published scientific papers.

Mandelbrot was born in Poland and spent most of his adult life in the United States. He launched the sciences of fractal geometry and fractal statistics in the last quarter of the twentieth century and coined the term fractal to describe a motif that repeats itself on ever diminishing scales. Fractal is a geometrical concept and refers to any geometric property of an object that reproduces itself at every scale. Think of nested Russian dolls, babushka or matryoshka dolls, with a never ending telescoping downwards of a peasant girl to smaller and smaller figures. A set of real matryoshka dolls consists of figures which separate, removing the top reveals a smaller figure of the same kind inside, which has, in turn, another figure inside of it, and so on. The self-similar nature of a fractal object turns the traditional notion of measurement on its head since no matter how small the measuring instrument there is always a level of detail that is being left out of any measurement. To understand how the fractal concept influences observables consider the concept of a curve having a diverging length as was first investigated by L.R. Richardson (1881-1953), see Fig. (**3.20**).

Fig. (3.20). L.F. Richardson (1881-1953) on the left was the first to quantify the way in which the physical measurement of irregular curves, such as coastlines, diverge. B.B. Mandelbrot (1924-2010) on the right invented the concept of fractal to characterize such pheomenona and proved that the divergence observed by Richadson was a generic property of fractal objects.

So what is a fractal?

For a long time it was believed that the motion of an object is either described by a smooth regular orbit or trajectory, such as those of the planets and their satellites, or if not regular, then it is erratic and an ensemble of such erratic trajectories is described by a Normal distribution function. However what is apparent is often not true. As we said in the context of the three-body problem an orbit may be so irregular, see Fig. (**3.18**), that it cannot be described by the traditional kinds of curves. If a body deposits a trail as it moves, such as the ink flowing from a moving pen, would the trail be a simple line? Could the trail cover a surface? Could the trail fill up a volume? How can the dimension of a trail be related to the motion of a body? All these things depend to a large extent on the dimension of the space in which the motion is taking place and how much of that space is used up by the moving process.

3.4.5. Fractal Dimension

Scientists are accustomed to thinking about space in integer dimension. A point has zero dimensions; a line has one dimension; a surface has two dimension and a volume has three dimensions. Space with a dimension lying between the integers is a relatively new concept in science. Even with the introduction of statistics we still consider phenomena to have integer dimensions. For example, the statistical processes described by random walks, homogeneously occupy the available Euclid's space of one or higher dimensions. However, there are types of random walks that skip over points in space, resulting in discontinuous processes and ultimately the trail left by such an exotic random walker does not fill the available space, but leaves holes in it; think swiss cheese. The dimensionality of such apparently exotic processes, since they occupy only a fraction of the available space, are non-integer.

Montroll, along with his post doctoral researchers Michael Shlesinger and Barry Hughes, systematically incorporated the fractal idea into the behavior of random walkers, extending and developing some of the early ideas of Mandelbrot. If a fractal random walk takes place on a plane, the statistical process has a dimension that is greater than one, the dimension of a line, but less than two, the dimension of the plane. In a similar way an irregular surface can have a dimension greater than two, but less than three. The surface of the mammalian lung bursts from the confines of two dimensions to nearly fill the third dimension, with an area confined to the chest cavity that is so irregular that when it is smoothed it can cover a tennis court. The growth of a lung, from a bud to a fractal structure, with a dimension between two and three implies that the code controlling that growth may have an algorithm that generates a fractal structure, as discussed elsewhere [46].

It is worth emphasizing that scale and dimension are intimately related; if a process can be characterized by a definite scale, then it has an integer dimension. This is true whether the process is deterministic or stochastic. On the other hand, if a process does not possess a characteristic scale, but perhaps many scales are present, then such a process often has a non-integer dimension. We refer to this non-integer dimension as a fractal dimension. I can assure you that this simplistic explanation of a non-integer dimension would send a mathematician running from the room as if his hair was on fire, but it is sufficient for our purposes.

The geometry of many objects is determined by the scale of a measuring instrument, for example, the length of a fractal curve is determined by the size of the ruler used to measure it. This sounds more metaphysical than scientific, so we shall spend some time clarifying the issue. After all is said and done, the length of a curve is what it is. The length of the curve has objective reality. Doesn't it?

In Fig. (**3.21**) a curve with many twists and turns is depicted. The curve was constructed from a mathematical expression that insures it has a fractal dimension. It is clear that using rulers of different sizes, as indicated in each sketch by multiple short line segments, yields a specific length of the curve determined by how many times the line segment is applied to the curve end-to- end. It is clear that applying the ruler ignores a great deal of structure smaller than the length of the ruler. In the middle sketch, the length of the ruler is one-quarter the length of that used in the top sketch. Therefore the total measured length of the curve is found to increase, but not by a factor of four. The increase in the measured length is due to the additional small scale structure that can contribute to the measured overall length using a smaller ruler. The bottom sketch indicates that reducing the ruler by a factor of 20 from the top panel, again increases the measured length of the curve. On the scale of the image in the last sketch it appears that the measurement has captured all the structure of the curve. However if the image of the last graph were magnified it would be clear that even smaller structure remains that is have not taken into account. A fractal curve does not have a characteristic scale, the measured length of the curve becomes longer and longer, eventually becoming infinitely long as the size of the ruler shrinks to zero. Precisely how this divergence occurs depends on the fractal dimension of the curve being measured.

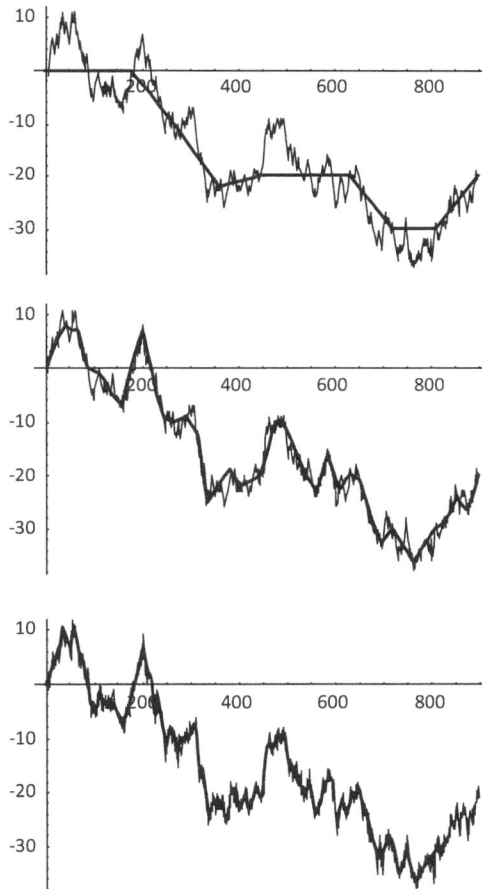

Fig. (3.21). The length of a mathematically generated fractal curve is measured using rulers of three different lengths. With each ruler we obtain a different length of the curve. Consequently, if the curve continues to have structure at all scales the traditional notion of length collapses.

Here we see the almost cabalistic result that a line can be drawn between two points a finite distance apart that is so irregular that it is infinitely long. This is where the lack of linearity makes the world seem eerie. We have this image of measuring a thing, say a shirt or dress, and the size is subsequently a fixed quantity. Operationally, if a ruler of a given length, 1 inch say, is laid end to end twenty-four times, then the length of the object we have measured is twenty-four inches long. Furthermore, we believe that this length of twenty-four inches is an objective property of the thing measured. The object has this length, or one very close to it, no matter how we choose to measure it. This cognitive picture of the world works very well as long as we do not encounter an object that is too irregular. Or does it?

The counter-intuitive idea of having infinitely long curves connecting points a finite Euclidean distance apart, was introduced into the physical sciences by the remarkable scientist Richardson [62]. He was a British physicist and Quaker, interested in the causes of war from the perspective of avoiding them. His early career was in physics, where he was the first to introduce numerical methods into the prediction of the weather for which he was inducted into the Royal Society. He subsequently left physics and focused his attention on problems in the social arena. His research led him to the unexpected and, from the view of many social scientists, unwelcome conclusion that quite independently of the traditional reasons generally given to justify war, the number of wars initiated globally in any given interval of time follows a bell-shaped curve, not the Normal distribution, but one very close to it and invented by Poisson.

Richardson used all the available data to count the number of years in which no war began, one war began and so on. His results revealed that the number of wars could be well represented by the one-humped Poisson distribution. The Poisson distribution describes a process in which events occur in a random manner independently of each other. The probability of a war being initiated within any time interval is proportional to the duration of the interval. We do not pursue this example further, but mention it only to demonstrate that Richardson was a person to be taken seriously and once embarking on a course of research he followed where ever it took him.

In his effort to understand the causes of war, Richardson began to consider border disputes, as one such cause. He examined how the borders between countries were determined and in the length of coastlines. His common sense idea for measuring the length of a rugged coastline was to place a continuous string along the contour of the coastline in an Atlas, measure the string's length and assign that as the length of the coastline. However, if another continuous string is used, but one with more flexibility and of smaller diameter, the string will fit into more crevices of the coastline and another number for the length of the coastline is obtained. The length of the coast therefore depends on the properties of the measuring instrument. This was the insight that Mandelbrot used to birth the notion of fractals.

The historical concepts of length and dimension were based on the assumption that the world is fundamentally a linear place. Even in the case of the erratic curve just discussed. A reasonable person would implicitly assume that when any given wiggle of the curve is sufficiently magnified it will appear smooth. At this level of resolution a linear ruler can be laid down end to end to measure the length of the curve. Any smaller ruler just reproduces the previous result with more placements along the length of the curve. Consequently the fact that no such limit exists in fractal curves implies that fractal phenomena are not linear.

3.5. THE TYRANNY OF EXTREMES

When my sons were in high school their sport of choice was wrestling. Every parent who has watched their child wrestle comes away with a sore back, legs that go into spasm, or any of a number of other tension-induced muscle cramps. As parents we know that an arm can be stretched only so far and then a muscle tears, a tendon pulls free, or a bone breaks. Any or all of these extreme outcomes play through a parent's head in the course of each of the three minute segments of a match. It is in trying to remotely pull our child's arm free that we dislocate our own shoulder, but to no avail. The wrestling match is dominated by our empathy and if the rare event of a bone breaking occurs the match would be stopped. That is the nature of extremes, they qualitatively transform the activity.

3.5.1. Tipping Points

My concern over the breaking of bones is not just macabre fantasy; someone I know had the chain on their bike come loose while they were pedalling down hill; another person's family car had their brakes fail; a favorite chair collapsed while a friend was reading a book; in short, myself and everyone else in the real world has experienced that mechanical devices and structures have critical points beyond which they no longer function as designed; they fail. The failure of a mechanical structure is obvious, but more subtle failures occur in the washout of a relationship; the disruption of a communications; the collapse of a social contract and many other non-physical structures. The rare event that unexpectedly changes the familiar and comfortable into something totally different is today called a tipping point, a term the talented author Gladwell popularized in a book by the same name [63]. What does a tipping point imply?

A tipping point stands out most clearly when there are only two states; black or white, yes or no, up or down, for or against, and the slightest touch or smallest piece of information induces an unexpected transition between the states. Consequently tipping points are uncertain and surprising. They can be described after the fact, but they are not predictable, or at least they have not been up until now. The inability to predict when a tipping point will be activated is not the same as not knowing it exists. I can know that a situation has come to a head and still not know how to predict the outcome.

The surprise aspect of tipping is like that of getting a joke, or having an insight, the *Eureka* moment that propels you naked down a public street announcing your discovery at the top of your lungs. There is no linear extrapolation to indicate that after thinking for a given length of time, understanding will suddenly emerge. The insight is unexpected and its source is unknown, but the process is certainly not linear. Sometimes it comes by retracing familiar unsuccessful arguments that ultimately reveal a hidden contradiction or inconsistency. A contradiction is desirable, because the truth lies on one side or the other and hence the contradiction reduces the number of options to two. In the face of uncertainty this is as good as it gets. At other times insight comes by abandoning all the traditional approaches and trying something that on its face cannot work. This last approach is not for the faint of heart and I do not recommend it for those who take personal criticism seriously. I have found that most people are all too willing to explain why a speculative proposition will not work, cannot work, and how such thinking is suspect for even suggesting it.

However things are rarely so simple as having two choices. Even in the world of politics where the choice is between two candidates, this is not always the case. An election can have a doubtful outcome and one side or the other cries foul. Some Democrats still claim that Bush stole the 2000 presidential election from Gore in Florida; a significant number of African Americans maintain that slavery did not end with the Civil War; and many Irish are quick to anger over the 700 year occupation by the British. The number of examples is limited only by imagination where the choices appear to be two, but in fact the controversy contains many more sides.

The threshold, the tipping point, the final straw, the game changer, are all slightly nuanced catch phrases for configurations that have an enhanced sensitivity. One push and the car goes over the cliff, the wall topples, or the armed sociopath fires his weapon into the crowd. At the tipping point survival often requires an ability to adapt and such ability is in all likelihood a consequence of anticipating the 200-year earthquake, or the 100-year flood, and making the appropriate preparations.

In the simplest example I tip over a glass of red wine, spilling it irretrievably onto the white rug. In a slightly more complicated case I publish a book and for some reason or set of reasons it becomes a best seller. (You can tell this is a fanciful example.) There is a transition from one state to another state across a barrier that separates them. Toppling the wine glass is simple and can be done either intentionally or by accident, with the same result. An essential feature is that it cannot be reversed. The wine cannot be removed from the rug and put back into the glass. By the same token transforming a book from a door stop into a best seller is not under my conscious control or that of any other single person. Here again the irreversibility of the tipping point is evident, once the selling begins it cannot be stopped. It becomes an extreme event.

A question that arises is whether a given process has a single tipping point, or are these points rare, unexpected and infrequent, but not unique? Can there be multiple tipping points? If the underlying process is random with Normal statistics it has been determined that extreme events have a statistics all their own and with each extrema it is possible that the process can be transformed or tipped.

When I was living in San Diego I would frequently swim in the ocean and I always answered two questions before taking the plunge. What is the temperature of the water? How high are the waves? In both cases the higher the better; within reason. I did not expect the variation in temperature to be very great while I swam, perhaps ten degrees, if I passed through a cold current. On the other hand, the maximum wave height might vary up to 25% over an hour or so. These are unscientific estimates I conjured up from my own experience, for the sake of example. The point is we do this kind of conjuring all the time; estimating changes and making projections, often from little data. In the previous chapter I used ocean waves as an example of a complicated phenomenon that for many purposes may be considered to have Normal statistics. These assumptions implied that for wind generated water waves that are not too high an oil spill behaves in a predictable way. The oil disperses over the ocean surface like milk diffusing in my morning coffee. What I did not discuss was how this would change if I stirred the coffee, or if a sudden storm mixes the surface waves. In a stormy sea the wind-generated waves are blown over onto themselves, they become unstable and the surface is dominated by breaking waves and ocean spray. This extreme behavior does not occur in a uniform and predictable way, but just as a typical wave field obeys the law of error, so too, the extreme values for breaking waves obey laws. However, even when the statistics of the underlying phenomenon are Normal, those of extreme events are not.

3.5.2. Simple Extrema

In the middle of the last century a German mathematician Emil J. Gumbel (1891-1966) developed a new distribution that describes the behavior of rare events in stochastic phenomena, which now bears his name. The Gumbel distribution has been applied to a wide variety of extreme events including wind gust loads on airplanes in flight, highest temperature or lowest pressure in meteorology, flood and drought in hydrology, and human life spans. To be prepared for the rare, but inevitable occurrence of extrema, planners must understand the genesis of such events. This distribution is based on the assumptions that data detailing the phenomenon of interest is of the Normal or exponential type and the underlying stochastic process that determines the magnitude of the variable of interest does not change over time. The Gumbel distribution therefore characterizes rare events for processes with Normal or exponential statistics. I refer to these jointly as exponential processes and consider an example of such an extrema process.

While I was a post doctoral researcher with Elliott Montroll, he suggested I study the 1300 years of data for the maximum and minimum height of the Nile River. At that time the best predictive model of the height of the Nile floods was a mathematical relation correlating the river height with the temperatures of Dutch Harbor, Alaska and Samoa in the Pacific, and the barometric pressure at Port Darwin, Australia. As might be expected, the predictive lifetime of this relation was not exceptionally long [64].

The Nile River is one of the longest rivers in the world, measuring 4157 miles from its most distant source to the entry of the Rosetta Branch in the Mediterranean. Nile River flooding was extremely important to the Egyptian people. The main Nile flows from Khartoum to Aswan in desert terrain, with only a narrow strip of vegetation covering each bank. From Aswan to Cairo, the Nile is bordered by a flood plain of alluvium, which gradually increased to a maximum width of twelve miles in the direction the river flows. Thus, long before Herodotus traveled through Egypt in the fifth century BC the people prayed for 16 ells. Pliny explained:

> "...12 ells means hunger, 13 sufficiency, 14 joy, 15 security, and 16 abundance."

It was therefore important to know the maximum height of the river in a given year. Budgor and West [64] explain that a network of twenty Nilometers was erected at different points between Aswan and Cairo to measure the river level. They were cylindrical marble columns placed in the water by the river banks, approximately 5.3 meters in height, 1.9 m in circumference at the base, and 1.6 m in circumference at the summit. On the north, south, southeast and east faces, a series of gradations were etched, each series being calibrated from a different zero-point. A fine degree of measuring accuracy was therefore obtainable [64]:

> "Two eagles, one of each sex, were placed at the summit of a Nilometer and on the first day of the flood, the Pharaoh, his priests, and the populace, deep in prayer, would listen for which of the two eagles screamed first: if it was the male, a great flood was predicted and Pharaoh immediately raised the price of the yet unsown corn. This is probably one of the first documented instances of religious exploitation of the people and the first practical argument for the separation of church and state."

Using 1300 consecutive years of reliable Nile River measurement data, it is a simple matter to test the applicability of the Gumbel distribution. In Fig. (3.22) the straight line segment corresponds to an exact fit of the transformed data to the Gumbel distribution. It is evident that throughout most of the heights the data are well fit by the straight line segment so that the statistics of the Nile River minima are probably described by an exponential process. We do not show the Nile River maximum heights because that was not as well fit by the Gumbel distribution alone as was the minima. This failure to track the river maxima suggests that the yearly river maxima were not independent of one another and consequently additional analysis was required to understand the flooding data [64].

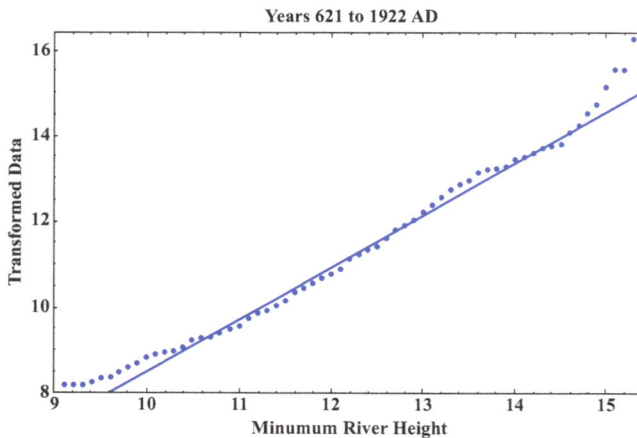

Fig. (3.22). The 1300 years of data from 621 to 1922 A.D. for the minimum height of the Nile River are transformed using the Gumbel distribution. The transformed data are plotted against the river height. The straight line would be an exact fit to the Gumbel distribution.

The ability to predict extrema whether it is the maximum or minimum of any given complex phenomenon has advantages for society. Some benefits are clear such as in the design of levees to hold back river waters in preparation of the 100-year flood. Others are perhaps more subtle, as in the anticipation of the maximum number of blackouts that may occur during the summer months in order to make preparation for the subsequent medical emergencies. How successfully city planners allocate resources to mitigate disasters is very much dependent on how well they are able to predict the occurrence of extrema.

The Planner's Dilemma is depicted in Fig. (**3.23**). As the magnitude of the event increases so does the cost of preparation. This increasing cost curve is relatively easy to construct since planners usually know the incremental preparation costs for an event of a given size. The more difficult curve to construct is the probability of the occurrence of a given size event. If the probability curve falls to zero very quickly, then the preparation costs are probably modest. They are modest because the size of the extrema within say a 100-year window is not too severe. However if the probability continues to have non-zero values far to the right, so that the probability of a very large magnitude event is high, the preparation cost might be prohibitive. An inverse power-law distribution could have very large extreme values occurring within a given time window.

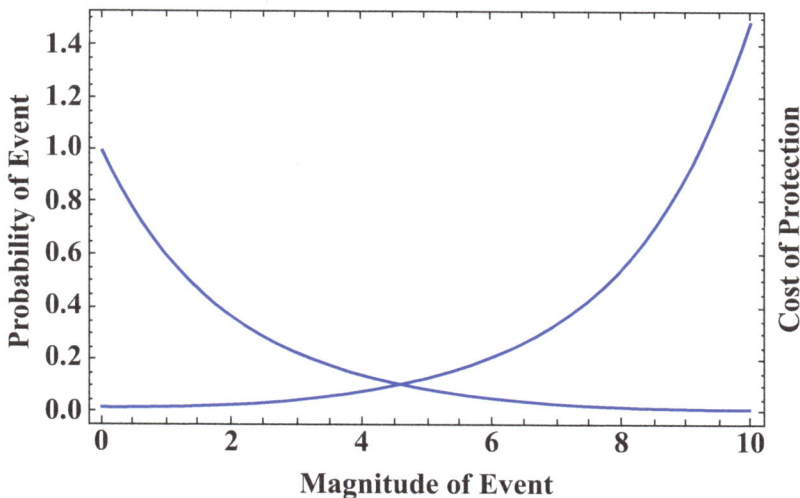

Fig. (3.23). The Planner's Dilemma: the cost increases as the probability of a rare event decreases, both as a function of the magnitude of the event.

The planner is faced with the prospect of spending an ever increasing amount of money to protect against an ever less likely disaster, such as a flood of a given height, or an earthquake of a given magnitude. The difficulty is how to balance the likelihood of a catastrophic event against the cost of mitigation against it. The only way to rationally construct such a trade-off between the pocket book and peace of mind is to have a systematic estimate of the probability for the occurrence of an event that may never have been measured. To accomplish this trade-off the planner uses the fact that the waiting time for an event of a given size is inversely related to the probability of an event of that size occurring. For example, the city might wait ten years for the occurrence of an event that had a 0.10 probability of occurring, but it would wait 100 years for an event whose probability was 0.01. In other words if the underlying process is exponential and the probability of an extreme event not occurring was 0.99, then typically the city would be safe for 100 years.

So what does the city planner do when the statistics of the underlying process is not exponential?

3.5.3. Complex Extrema

The mathematical study of the statistically unusual or rare event is known as extreme value theory. It is a branch of statistics that deals with extreme deviations of the probability density function from the median. Half the data is less than the median and half is greater, so that the median is not the same as the mean, except when the probability density is symmetric. Suppose I wanted to buy a home in an upscale neighborhood and the real estate agent quoted the selling price of the house I was interested in buying at a million dollars. My wife then goes on the computer and looks up houses that have been sold in the immediate neighborhood and finds that the average price of the houses sold is comparable to the quoted price. That is where most people stop the investigation, but not Sharon. She looks a little deeper by ordering the most recently sold homes according to their selling prices and finds that more than half the houses sold for less than half a million dollars. Therefore there are houses at the very high end that are skewing the average cost towards a higher price and consequently we would be paying a premium to live in the same neighborhood as those houses. Of course this is not the sole consideration in buying a house, but it is one that should be carefully considered.

When the mean exceeds the median the distribution is skewed toward the high end. When the mean is less than the median the distribution is skewed toward the low end. The two only coincide when the distribution is symmetric, as it is for Normal statistics. However for many complex phenomena the inverse power-law distribution has replaced the Normal, in fact, this replacement is often used to define complexity. Consequently in these complex situations the mean always exceeds the median due to the heavy tail in the distribution. The long tail implies that there are many more large magnitude events than there are in exponential processes. So how does this effect the distribution of extrema? Recall that one of the conditions for the proof of the central limit theorem was that the variance be finite. This particular condition is violated by many, if not all, empirical inverse power-law distributions. With the violation of this condition the Gumbel distribution ceases to represent the behavior of the extrema and more general forms of extreme value distributions are required.

Leonard Tippett (1902–1985) invented the field of extreme value statistics. He worked for the British Cotton Industry Research Association, and did research on understanding how to make cotton thread stronger. His studies verified that the strength of a thread was controlled by the strength of its weakest fibers. The research determined that the most intuitive comparison between extrema is given by the return time, that is, after a given magnitude event occurs how long will it be before another event of that magnitude occurs? In this way the statistics of the size of an event can be predicted from the variability in the size over time.

The inverse of the probability of occurrence of an event of a given magnitude is plotted in Fig. (**3.24**) *versus* the return time. It is clear from this figure that the return time for inverse power-law phenomena is significantly shorter than that for exponential phenomena. For a Gumbel distribution, which describes the extrema statistics for an exponential process, a magnitude 50 event occurs every 100 years, but for an inverse power-law process this event occurs every 10 years or so. It is evident that it is much easier to develop protection against an exponential process than for the significantly more frequent inverse power-law process. This explains, in part, why levees built to withstand the 100-year flood are successful; the size of the rainfall contributing to floods are exponential processes. On the other hand, the magnitude of earthquakes are inverse power law.

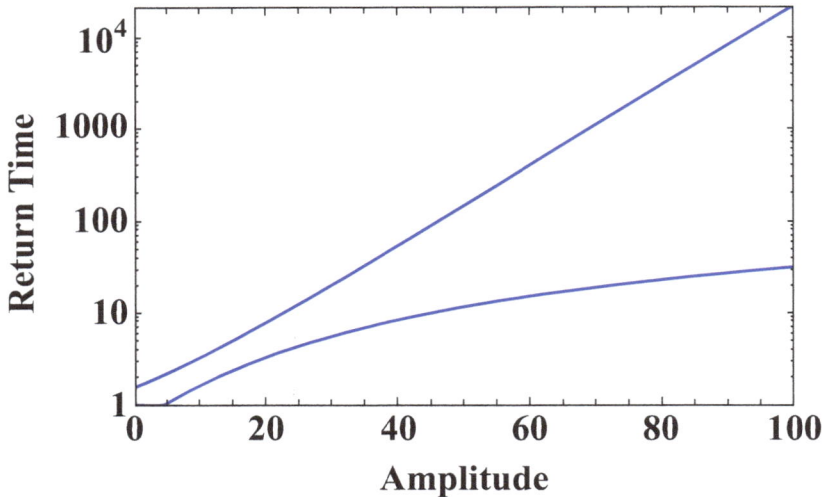

Fig. (3.24). The upper curve is the return time for a given amplitude event when the underlying process is exponential and corresponds to a Gumbel distribution. The lower curve is the return time for a given amplitude event when the underlying process is inverse power law.

You may recall my pointing out that very often whether a process is linear or not depends on the representation one uses. A similar observation can be made in the analysis of extrema. In the present comparison of the exponential and inverse power-law phenomena it is possible to prove that the two are simply related. The asymptotic inverse power-law distribution can be obtained from the exponential asymptotic distribution by a logarithmic transformation. What this means is that the return time for the inverse power-law data can be estimated from a Gumbel distribution expressed in terms of the logarithm of the original data. Or said differently, the lower curve in Fig. (**3.24**) can be transformed to the linear form of the upper curve by plotting the return time against the logarithm of the amplitude of the event. Of course all the numbers on the axis will be different but the functional form will be that of Gumbel. This is one of those tricks that scientists are fond of using that all too often merely confuse the uninitiated.

What I have attempted to bring out in these last few pages is that phenomena with exponential statistics are fairly straight forward to understand and their extrema are rare and relatively easy to prepare for. On the other hand, phenomena with inverse power-law statistics, those corresponding to complex processes, are difficult to understand and their extrema are much more frequent. It is these complex extrema that require special consideration and major investment of resources. Unfortunately when complex phenomena are mistakenly described using exponential statistics society is ill prepared for the extrema when they come. This is what happened repeatedly in the stock market in the twentieth century and each time a flash crash occurred it was treated as an anomaly. Such events ought to have been expected, but they never were, even though they are an intrinsic part of stock market dynamics [60]. It is also becoming apparent that other social phenomena suffer from the same shortsightedness on the part of the planners, such as in health care systems.

3.6. HOW UNFAIR IS IT?

In the last chapter I argued that the social concept of fairness is a consequence of the linear additive world view of Gauss, Adrian and Laplace. Moreover, that fairness is a consequence of having a well defined average that characterizes people and their activities, with the goal of society being to reduce the variations in what people are paid, the variability in what they are allowed to do and to insure that individuals are not 'exploited'. In the law this takes the form of protecting an individual's right to not be offended in the work place, to be sensitive to another's point of view, to insure that society becomes homogeneous with regard to race, religion, age, national origin and so on. The most desirable outcome for society is one that has maximum diversity, with the least degree of heterogeneity. Regression to the mean is not an idle phrase, but is instead the goal of virtually every Utopian dream. It is true that everyone flourishes in Utopia, but the point is that no one does not and therefore there is little to distinguish the king from the town drunk; they are of equal social merit. But then does anyone drink in Utopia?

In the real world inequities exist and the direction of civilization for the past two thousand years or so has not been the steady eradication of these inequities. It has been more of an incoherent dance this way and that, crushing some imbalance, endorsing others and even putting new ones in place; only to return at a later time and eliminate them again. Part of the reason for the dance is that some inequities are created and imposed by humans, whereas others are intrinsic to phenomenon and have a natural origin that emerges from complexity. The question was and remains how to tell the two apart.

When is a natural imbalance exploited for the benefit of the few? Some would argue that capitalism, with its imbalance in the distribution of wealth is intrinsically unfair and I would agree. A more important question is whether the inverse power-law distribution of wealth is necessary for a stable economy, since if the economy crashes no one will have any wealth, except perhaps the strongest and most vicious. Putting the moral questions, such as those of Malthus aside, it would appear that the inverse power law arises as a mechanism to control stability as a society becomes more complex. As the individual craftsman is replaced by the factory worker, the process by which the manufactured goods are distributed within society markedly changes. Modern society is not a scaled up version of the medieval one, cities transformed how humans think and behave so they no longer coincide with what they were in an agraian age.

As mentioned elsewhere [46], the phenomenon of writing, talking, biological evolution, urban growth, and making a fortune are more similar to one another statistically, than they are to the heights of individuals in a large population. The chance of a randomly selected child from a large group, becoming rich and/or famous, as they grow older, is not very large, but it is still much greater than that random child becoming tall. Part of this good fortune is the difference in the intrinsic nature of the underlying process.

Thus, the data on wealth, language, urban growth, biological evolution and apparently most other complex social/biological phenomena have distributions that are more like the inverse power law, than they are the bell-shaped curve of the Normal. One consequence of this difference is the imbalance in the underlying process. Why should a few percent of the population have the largest fraction of the wealth? For convenience suppose that 20% of the population owns 80% of the wealth. Of course less than 20% of the population owns more than 80% of the wealth, but historically this metaphor has been used to discuss the imbalance implied by the inverse power-law distribution and I continue in that tradition. The imbalance between the rich and the poor is determined by the Pareto index. The actual numerical value of the index is not important here; it is the existence of the imbalance within the data that is important. The 80/20 Rule has been called Pareto's Law in which the "few (20%) are vital and the many (80%) are trivial".

The science fiction writer Theodore Sturgeon independently identified the 80/20 rule in a form that in literary/artistic circles became known as Sturgeon's Law [65]:

> "Using the same standards that categorize 90% of science fiction as trash, crud, or crap, it can be argued that 90% of film, literature, consumer goods, *etc.* are crap. In other words, the claim (or fact) that 90% of science fiction is crap is ultimately uninformative, because science fiction conforms to the same trends of quality as all other artforms."

Of course it is not just the distribution and application of human talent that manifests such imbalance. In a large study [66] of radical Jihadist forums, the investigator Awan determined that 81% of users had never posted on the forums, 13% had posted at least once, 5% had posted 50 or more times, and only 1% had posted 500 or more times. This may be interpreted as an 81/19 imbalance between those that participate in the Jihadist forums on the internet and those that merely observe what is being done. On a Wikipedia site [67] this observation was linked to the 1% Rule that states that the number of people who create content on the internet represents approximately 1% (or less) of the people actually viewing that content. This means that for every person posting on a forum, there are at least ninety-nine others viewing that forum but not posting. The intellectual wealth is generated by the few and it is passively absorbed, but actively used by the many.

A number of business people have pointed out that 20% of the people involved in a project produce 80% of all the results; that 80% of all the interruptions in a meeting come from 20% of the people; resolving 20% of the issues can solve 80% of the problems; that 20% of one's results require 80% of one's effort; and on and on and on. Much of this is recorded in Richard Koch's *The 80/20 Principle* [68]. The 80/20 Rule works both positively and negatively in the business world. It is probably worth pointing out that it is not always the same part of the work force that fits into these percentages. For example, the 20% that is providing the lion's share of production has small overlap with the 20% that is causing most of the problems.

The value of Pareto's Law is that it is a constant reminder to focus on the 20% that matters. Only 20% of what people do during the day actually matters. Correspondingly 20% of the workforce produces 80% of the results that are achieved. Identify and focus on those things and those people. When the daily minutia consume your time, try not to get overwhelmed and remind yourself of the 20% on which to focus. If something isn't going to get done, make sure it's part of the 80%.

In corporations Pareto's imbalance in the distribution of achievement translates into the majority of the accomplishments being attained by the minority of the employees. It is the vital few that have the talent and ability necessary to generate revolutionary ideas and carry them to fruition. These are the intellectual leaders that focus the attention and energy of the majority on the tasks required to produce results. It does not matter whether these results are new ways to present and promote products to the customers, or they are new chemical formulae resulting in a totally different line of products. It is true that new methods and products are sometimes produced by one of the many, this is the exception. Such an individual is often catapulted into the ranks of the few. In any large group it is the 20% that produce most of the results, but it is not always the same 20%. There is a flux from the many to the few and back again and only the rarest of individuals remains permanently within the top ranks.

On the bottom end, in any organization, it is possible to identify a small group of individuals that produce most of the interruptions at organizational meeting. The same individuals ask the embarrassing questions that are discussed over lunch; the questions that make the boss uncomfortable, or open the floor to conversations that management would have postponed until they arrived at some decision before going public. It often seems that the 20% that produce 80% of the interruptions, seem to know that the new travel restrictions are being violated by the company president; that the new local policy for sick leave is inconsistent with that in the corporate office; and they certainly know all the office rumors.

The problems inhibiting the success of an organization often appear to be mutually independent, but in point of fact, they often are not. Problems, like results, are usually interconnected, however subtly, so that generating solutions is very similar to generating results. In practice this means that certain solutions are connected to multiple problems within an organization; the distribution of solutions has the same imbalance in their influence on problems as individual talents have on the distribution of income among people. This imbalance is manifest in 80% of the problems being resolved by 20% of the solutions, the other 80% of the solutions being sterile. Most proposed solutions do not have a substantial impact on an organization's problems and their implementation are costly in the sense that the cost per problem resolution is approximately the same for all solutions generated. Consequently, it would be highly beneficial to determine if a given solution is one of the 20% before it is implemented and thereby markedly reduce the cost associated with problem solving.

As a final example, the majority of the activities of most people do not produce significant results, that is, results of the 20% kind; whether the activity is research, problem solving or even painting. That reminds me of a story I often tell regarding the artist Pablo Picasso. It is one of those stories that should be true even if it is not. As you know, Picasso was a prolific artist and many copied his various styles. An acquaintance of his brought a painting to him and asked if he had painted it, to which Picasso replied: "It's a fake". Being unsure of this response the person went back to the art dealer from whom he had bought the painting. The dealer verified that the painting was, in fact, an original Picasso. The next day the person went back to the artist and confronted him with the claim of the art dealer, to which Picasso replied: "Yes, I painted it. I often paint fakes".

The point of this story is that Picasso realized that not every painting he did was a masterpiece, and if it was not a masterpiece, in his eyes it was a fake. Consequently, a significant fraction of Picasso's time was spent in producing fakes; perhaps not the 90% of Sturgeon's Law or the 80% of the Pareto Principle, but some large fraction of his time. The same is true for mere mortals like you and me. If left unchecked, most of what we do would not be significant and to change this imbalance requires a vigilant conscious effort of will.

CHAPTER 4

Unequal: A Matter of Scale

Abstract: The transition of our mental models from a simple to a complex world view, entails the breakdown of Normalcy and the necessary adoption of Pareto's inverse power-law distribution. The complexity measure in this new world view is the inverse power law index, whose magnitude determines whether or not variability of the underlying process can be described by a finite variance. It is often the case that in such phenomena the focus shifts away from continuous dynamics of mechanical systems, such as the trajectory of a person's life, to the time intervals between discrete events, such as having a heart attack or receiving a message. This shifting is particularly evident in information-dominated systems, whose time series may not even possess an average time between events. The appropriate quantities to measure in such fractal dynamical systems are not easy to identify, in fact, what we choose to measure may well be determined by how we define information and how that information changes in time. How information flows in complex networks, or how information moves back and forth between two or more complex networks, is of fundamental importance in understanding how such networks or networks-of-networks operate. This information variability is determined by the inverse power-law distributions, which in turn are generated by a number of generic mechanisms that couple contributing scales together. We identify different mechanisms that produce empirically observed variability; each one prescribing how the scales in the underlying process are interrelated.

Keywords: Allometry, Contagion, Criticality, Decision making, Frequency, Inequality, Inverse power laws, Networks, Rank order, Scaling mechanisms, Space, Time, Universality.

Science has historically been concerned with the search for universal principles and laws to describe and understand the world; relationships that capture wide ranges of experimental results with a minimum number of assumptions. This has been particularly true in the physical sciences. Sometimes a physical law is straightforward, as is the first law of thermodynamics; the conservation of total energy. But once the law is articulated, its implications are remarkable in the extreme. Conservation of energy implies that changes in kinetic energy are compensated by changes in potential energy and the exchange of one form of energy for another, such as in a swinging pendulum, provides a way to understand the dynamics of the mechanical part of the world in which we live. The clockwork universe of Newton, shown by Hamilton to be based on this conservation law, satisfactorily explained how things work for a large fraction of the civilized world. But others, who were more qualitatively minded, maintained that this was not adequate for characterizing the dynamics between people; it did not and does not encompass the human sciences. They wanted something more.

As a scientist I try to represent phenomena in simple forms, or some might think in simplistic ways. I think about such physical concepts as space and time in terms of how many times I can lay a ruler end to end to measure the dimensions of my room, while time is measured in terms of the number of ticks of the clock sitting on my desk. However, as a person, the dimensions of my room are much more than the number of square feet enclosed by the walls. The size of the room is also determined by the books that fill the shelves along each wall; the light shining through the windows overlooking the front lawn; and the connection to the outside world through my computer. Time goes faster or slower depending on whether an exciting insight is revealing itself in writing the right word or phrase, to explain a central idea, or is being stubbornly elusive. In this chapter I discuss the quality of life experienced in personal space/time, in terms of operational definitions introduced in the sociophysics of the nineteenth century. This discussion provides some of the insights necessary to understand the source of some of the inequality in the human realm.

Musicians and physical scientists understand how the vibrations of a violin are produced, but the aesthetic appeal of music is put into a category outside the science of acoustics. The human sciences of psychology and sociology lack the universality of physics: symmetry principles and the laws of thermodynamics have no human science analogs, with the possible exception of information flow. This lack of universality arises, in part, because the appropriate metrics are lacking, that is, scientists do not understand what they ought to be measuring, or whether what they believe is important in a given context can be measured. In a physical interaction there is always the exchange of something tangible: momentum, energy, matter, or the element of an appropriate field, such as a photon or phonon. However, in a social interaction the far more elusive quantity of information is exchanged. Of course the intended message of the sender, contained in the words used to construct the message, can be very different from the understanding of the message, as interpreted by the receiver. In the real world the history of the sender and receiver are involved in the formation and interpretation of messages, including such things as posture, hand movements, facial expression, choice of words and so on. Note that this is not the information exchange proposed by Claude Shannon shortly after the Second World War. His working hypothesis abstracted sender and receiver to featureless points that contribute nothing to the message. All the human qualities mentioned, as well as others, are important in determining the information transmitted in a social interaction, but these are explicitly excluded from Shannon's definition of information. However, the notion of information used herein is more general than that postulated by Shannon over a half century ago and which is still used by communications engineers.

Consider how little information is transferred in terms of words used when, for example, a youth wants to impress a member of the opposite sex. Compare the information content of the words to the information contained in an awkward stance, a nervous tick, an inappropriate laugh, and the myriad of other uncontrollable things, that despite their best efforts, emerge in the exchange. Only in literature do words: outweigh how people carry themselves; impress more than how a person is dressed; or focus the attention more than the warmth of a smile. So how is this collection of indicators about a complex person transmitted and interpreted by another person. These concerns replace the simplifying assumptions made about the nature of information by Shannon, which started as his working hypothesis over a half century ago and are, for the most part, still used today.

I have been asked by a number of people in the course of my research,including the Directors of the Army Research Office and the Army Research Laboratory, why the elaborate formal methods of statistical physics are not sufficient to describe the dynamics of complex social and psychological networks? Moreover they wanted to know why these hallowed techniques do not determine the information exchange among complex networks and the flow of information within a complex network? At different times I have given different answers to these questions. When pessimistic, I argued that the methods cannot be applied, because the social and psychological networks do not satisfy the basic assumptions made for the physical techniques. The physical world is solid and tangible, whereas the most significant aspects of social interactions are intangible and therefore their measures are more difficult to quantify. When more optimistic, I answer that, yes, they can be applied, if we frame questions in the appropriate way, using such concepts as entropy. Finally, I often respond that sometimes these statistical methods can be applied, but not always. I have tried to be forthcoming in my response up the chain of command.

An answer to the question of how to describe the transfer of information among complex networks, that is probably the most practical, was given by in the middle of the last century by the mathematician Norbert Wiener. Wiener was the first scientist to identify the intrinsic complexity in communications between and in the control of living organisms, in machines and in the interaction, between the two. He proposed mathematical methods for the description of their dynamics with the creation of the discipline of Cybernetics; the study of how to control the man-machine interface. The depth of his insight into the interaction between complex networks is revealed in his intuitive observation [69]:

> ". . . We have a system of high energy coupled to a message low in energy, but extremely high in amount of information, *i.e.*, of great negative entropy. This is unlike the usual interaction in thermodynamics, where all the coupled systems enjoy high entropy. But it may happen in the development of such a system that the internal coupling causes the information, or negative entropy, to pass from the part at low energy to the part at high energy so as to organize a system of vastly greater energy than that of the present instantaneous input. . ."

Recall that the second law of thermodynamics states: "heat flows from an object with the higher temperature, to an object with the lower temperature, when the two are in physical contact". Physicists used this mundane observation to deduce that the entropy of a closed system can only increase or stay the same over time, it cannot decrease. Eventually the entropy increase destroys the order within systems, through the erasure of memory, resulting in the prediction that the universe shall experience a 'heat death' after a very long time.

Wiener's speculation on the transfer of information between complex networks, appears to imply that the second law can be overturned when two or more complex networks, with differing information levels, talk to one another. But the second law is not violated. It is true that the network with the lesser energy, but greater information, can, under certain conditions, organize the network of lesser information, but greater energy. Influence is transferred from a complex network high in formation to one lower in information. The control follows the information gradient and works against the energy gradient. This is an information-dominated process. I named the direction of information transfer in this case the Wiener Rule, because it was the first attempt, albeit an informal one, at formulating a universal principle with which to guide the dynamics of complex networks. Wiener was, of course, speculating, and his speculation were not vindicated for another half century. The vindication required generalization of a number of statistical physics concepts, including the fluctuation-dissipation theorem [70, 71], whose physical interpretation will be given subsequently.

In this and the next chapter the importance of information, along with its transfer between complex networks, is discussed. However, the difference between energy-dominated networks and information-dominated networks, sets the stage for understanding why and how complex phenomena are described by the inverse power-laws introduced in the last chapter. In this way it is possible to quantify the level of network complexity using the inverse power-law index. This allows us to express the Wiener Rule in terms of the difference between the two indices characterizing the complexities of the networks.

The present-day understanding of Pareto's scale-free inverse power-law distributions in sociology began with small-world theory; a theory of social interactions in which social ties can be separated by the strength of the links between network members [72]. Strong ties exist within a family: parents, siblings, cousins and so on; among the closest of friends, those listed in emergency contact numbers, and called to tell about new babies, promotions and firings. Then there are the weak ties: such as between most fellow worker, with whom you chat, but never reveal anything of substance; friends of friends, and business acquaintances. It is also true that there are levels in each of these two categories; the closeness of a brother typically trumps that of a cousin, while the empathy with a scientific colleague eases out that with a random office mate.

Clusters form among individuals with strong interactions to form closely knit groups, small groups in which everyone knows everyone else. These tightly bound clusters would be isolated, if not for weak social contacts that couple such clusters together. Weak links connect from within a cluster, to other clusters within the larger network. The weak ties are all-important for determining the interaction with the world at large. This 'small world' has short cuts that allow for linkages between one tightly bound small group, to another tightly bound small group very far away. It is possible to link any two randomly chosen individuals, with a relatively short path across links using only a few of these long-range random connections. This has become known as the six-degrees of separation phenomenon. Any two individuals selected at random from the population of a large city or country are connected, on average, by five other people in the population, who are coupled to each other sequentially, through a chain of acquaintances, but are otherwise strangers. Consequently, strong short-range clustering and random, weak long-range connections are the two basic elements necessary for the small-world model.

Small-world theory was the forerunner of scale-free networks. Research into the study of how scale-free networks form and grow over time, reveals that even the smallest preference, introduced into the process of selection, has remarkable effects. We examine a number of such mechanisms, whose influence within a network has been shown to eventually lead to an inverse power law in some network property.

4.1. CRITICALITY AND MULTIPLE SCALES

The best days for snowball fights in Buffalo, where I grew up, were not the coldest days. The snow of choice was 'packing', meaning that it was moist, and could be compressed by hand into a compact sphere and hurled with accuracy at an unsuspecting opponent. If it was too cold, the old snow would freeze into ice, but the new snow would be wispy and virtually useless for making snowballs. Consequently, the relation between the phases of water were of great concern to an avid eight year old snowball hurler. Everyone in the neighborhood knew that, if pressed hard, a snowball would become watery, then partially freeze into the proper form, when again exposed to the cold. It would be years before I learned that compressing the snow, increased pressure and reduced the volume, thereby generating heat; heat raises the temperature, thereby breaking bonds between molecules, inducing a phase transition from solid snow to liquid water; by exposing the resulting wet snowball to sub-freezing air lowers its temperature, reestablishing the chemical bonds, and inducing a phase transition from liquid back to solid. The short-range interactions of molecules in gases, the intermediate interactions of molecules in liquids, the long-range interactions of molecules in solids and the phase transitions between these various states of matter, would eventually become part of my cognitive map of the physical world. It would take a good deal longer to appreciate that these phase transitions are all a matter of how multiple scales interact with one another.

One of the recurrent themes herein is variability, and its quantification using the variance of an empirical probability density or time series. In Normal statistics the variability is restricted, so that errors cannot become too large. For example, the variation in the normal length of the individual bones in the human body are not very large, resulting in a Normal distribution of heights in a human population. But this is not the only kind of variability. The Earth's changing weather can be measured in terms of the frequency and magnitude of storms. The variability of the occurrence of events in time is separate and distinct from the variability in event amplitude. Recall the two kinds of distributions in earthquakes; the Richter-Gutenberg law of quake magnitude is distinct from the Omori law for the time interval between quakes of a given magnitude. These two kinds of probability densities can be very different and yet they do not exhaust the kinds of variability people encounter every day. There is a third commonly used measure of variability that is again different from these two. The third measure determines the scale over which a given event influences subsequent events; this can be a correlation time or a correlation distance, the correlation in space or time being the measure of influence.

4.1.1. Time

Scaling links together the world views of Poincaré, with his nonlinear dynamics, and Pareto, with his inverse power laws. The term scaling might be a little confusing, because how it is interpreted has changed dramatically over time. The term scale itself has morphed from a noun to a verb. At the turn of the last century classical scaling related properties between successive generations. For example, a population size that increases by a constant factor from one generation to the next, produces geometric growth, just as it did for our dour friend Malthus. Classical scaling therefore refers to the constant scale factor that interrelates sequential generations. For example, Moor's law states that the number of transistors on a computer chip doubles (rate) every eighteen months (generation). In contemporary studies of complex phenomena, scaling no longer has the classical interpretation; today scaling implies that the phenomenon of interest does not have a characteristic scale. Earlier we discussed mathematical fractals, objects that do not have characteristic scales and now we examine some exemplars.

Consider playing a recording of violin music at various speeds. The distinctive violin sound is distorted and is eventually lost as the playing speed is changed. The fact that the singular sound of the violin disappears implies that the violin has specific vibrations or time scales that make its sound unique. It is not music that concerns us here, but the characteristic sounds of the violin itself, say those heard when an orchestra is tuning up. Of course, every musical instrument has its own set of characteristic frequencies and could be inserted into this experiment; the choice of violin is arbitrary. However there is a class of sounds whose quality is unlike that of musical instruments and is invariant to the speed at which it is played. These are scaling sounds, also called scaling or fractal noise.

Familiar examples of scaling noise are running water, static between stations on the radio and erratic snow seen on television when the signal is lost. Such scaling signals are called white noise, which is a colorless hiss having the property that a fluctuation (random audible frequency) at any given instant is independent of a fluctuation (random audible frequency) occurring at any other instant. Most familiar measures of the interdependence of the fluctuations at different points in time, such as a correlation function, are zero. Consider replacing sound with language, a language in which words are uncorrelated, that is, the words are independent of one another. Such 'verbal soup' characterizes certain types of mental disorder in which the patient can speak, but is incoherent, not unlike certain politicians and university administrators.

When I was a child my friends and I would sometimes play 33 1/3 rpm records on a 78 rpm turntable. Simply put, we played records recorded at one speed at a higher speed. This transformation of the sound invariably enhanced the higher frequencies, changing a crooner into a screechy soprano. We thought that was hilarious. Of course, we also found the complementary joke of playing 78 rpm records at 33 1/3 rpm and transforming a clear throated alto into a garbled bass equally funny. In this way I developed the intuition that when recordings are played faster than the speed for which they were designed their frequencies are shifted higher and the converse when they are played more slowly.

Imagine my surprise some years later when I was shown how certain kinds of fractal music when played faster actually emphasized low frequencies. What I observed with fractal music was that when the playing speed is doubled rather than the music being an octave higher it could be semi-tones lower, a phenomenon running counter to everything I had learned. Doubling the speed at which a recording is played is equivalent to changing the time scale and Schroeder [73] gives an elegant mathematical discussion of how this can result in a semitone lower in simple fractal music, rather than an octave higher as it would in traditional music. Consequently the simplest form of fractal music scales and is self-similar.

It is possible to develop some intuition of fractal time by listening to certain kinds of music. When I was teaching at the University of North Texas I designed a course for non-physics majors based on fractals. I did not give final exams, but I did require an original oral presentation at the end of the semester discussing one of the concepts introduced in the course. The university has a world class music school and some of my students, who were music majors, contributed original scores, composed using computer algorithms based on the mathematics of fractals. More than one of their compositions were truly beautiful.

I dwell on fractal music in part because such music is subsequently shown to have power-law spectra. What is interesting here is that an outside stimulus, such as changing the speed of a record, can lead to a counter-intuitive response. In a different context, such an unexpected response, rather than being humorous, could lead to unexpected and unwanted consequences. Consider a very different phenomenon, that of a prey-predator system in dynamic balance, with stable oscillations in the two populations, as well as, the predator maximum lagging behind the prey maximum in time. How would such a system respond to increasing the food supply for the predator, by increasing the population of prey. Wikipedia describes the phenomenon coined by Rosenzweig the *Paradox of Enrichment*:

> "He described an effect in six predator-prey models wherein increasing the food available to the prey caused the predator's population to destabilize. A common example is that if the food supply of a prey such as a rabbit is overabundant, its population will grow unbounded and cause the predator population (such as a lynx) to grow unsustainably large. This may result in a crash in the population of the predators and possibly lead to local eradication or even species extinction."

This is one of the intriguing aspects of working with models in science. They can lead an investigator into places that are not suggested by experiments and may even be counter indicated. When this happens a scientist is afforded the chance to determine just how well they understand the phenomenon by designing an experiment to test the theoretical prediction. Here we merely note, as did Lewis in his extraordinary *Book of Extremes* [73], that having too much of a good thing can result in bursting of a bubble, that is, the dynamic system can be destabilized and come crashing down.

4.1.2. Space

Today scaling in a dynamic process means that the response to stimulation at the smallest scale is not restricted to that scale, but influences the entire hierarchical structure. In a deterministic network, such as a piece of metal, this behavior is the result of the long-range nature of the coupling between the atoms of the solid. A small piece of solid cannot be moved without moving the entire object; unless of course a piece is cut off. Moving a small piece of fluid, on the other hand, is quite different. The parcel of fluid being moved only pulls along the fluid adjacent to it, by means of local interactions. This is interpreted as drag and is produced by the short-range nature of the forces acting between fluid particles. It is only in a gas that local motion is completely independent of what is taking place at other locations, since the interaction of gas particles is through localized collisions. Scaling is part of our everyday experience in which a process undergoes a phase transition: ice freezes overnight on the windshield of the car and melts when the car heater is turned on, both result from the change in the force law between particles. Such changes in the force law modifies the molecular coupling across scales.

Scientists have learned that phase transitions are not restricted to physical processes. In fact examples of this collective behavior are observed every day in schools of fish, flocks of birds, herds of various kinds and swarms of insects. The collective behavior of animals depicted in Fig (**4.1**) has intrigued people since the earliest recorded history. Questions immediately jump to mind on seeing such collectives. How do these groupings form and what holds them together? What controls the dynamics? Is there a leader or is the cooperation truly emergent behavior? Proposed answers to such questions range from collective consciousness to sophisticated mathematical models of local interactions and everything in between. A school of fish encounters a predator and rather than breaking apart and fleeing to survive, the school of fish instead acts as a unit, to maintain the illusion of size. The predator sees the size of the school, not that of the individual fish and is discouraged from attacking. It is clear in this situation that individuals do not act independently of the group; all activity is tied across scales. This collective behavior clearly has survival value for the group, so is this a consequence of biological evolution? Does nature favor those species that can cooperate? That strategy did not turn out so well for the buffalo that two hundred years ago covered the central plains of the United States and now are all but extinct. So is it that some collections have evolutionary advantage and others do not; or maybe some had advantage at one time but circumstances have changed? Some predators such as man are not fooled by the illusion of size, or at least not fooled for long.

Fig. (4.1). A flock of geese is depicted in the upper left panel; a school of fish in the upper right panel and a herd of buffalo in the bottom panel. All three are examples of cooperation in ecosystems.

An indication of how common these various assemblages are in nature is indicated by the variety of names we have given them; such as a bevy (of quail), a covey (of partridges), a clowder (of cats), flocks of all kinds, a gaggle (of geese on water), a gang (of elk), herds of all kinds, a kennel (of dogs), a plague (of locusts), a pride (of lions), a skein (of geese in flight), swarms of all kinds, and many more. Apparently experts want to capture the uniqueness of the group structure within a species by giving it a special name. This allows them to make slight adjustments in their mental representation of the natural history of the world. But perhaps more importantly these cognitive maps enable them to nuance the collective behavior observed in the social structures created by each of these species.

4.1.3. Decision Making Model

My friend and colleague, Professor Paolo Grigolini at the University of North Texas, together with a number of graduate students and post-doctoral researchers has developed a general mathematical model for the study of consensus. Their investigations suggest a new mechanism to explain the cooperative behavior seen in flocks, bevies, coveys, gangs and plagues. This grand decision making model developed with his students [74, 75] consists of a network of simple elements that have a decision to make, vote either yes or no, move either right or left, turn either on or off. Left alone each element randomly switches between these two states without bias as depicted in Fig. (**4.2**). However, when these individuals are allowed to interact with other members of the network in a way that depends on how many of their colleagues are in one state or the other, their behavior is dramatically changed. In fact there is a critical strength of interaction among the member of the network that induces a phase transition, very much like that seen in physical systems.

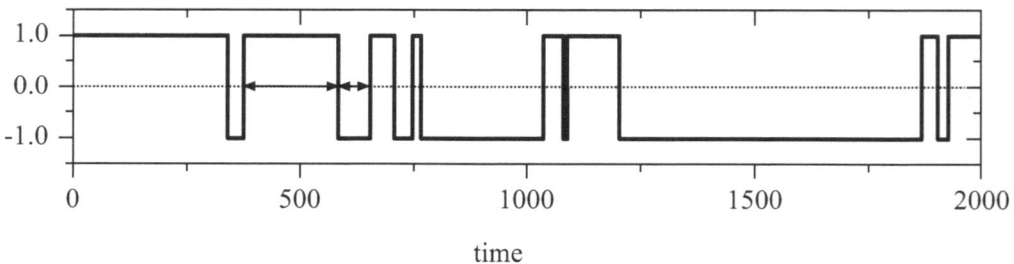

Fig. (4.2). A single isolated element in a decision making model [74] has an exponential distribution of switching times. The average time between switches is given by the rate in the Poisson distribution. [Adapted from Turalska *et al.* [75]].

A measure of the configuration of the network is the fraction of the difference in the number of elements in one state *versus* the other. When the interaction strength is sufficiently weak this fraction is depicted in Fig. (**4.3**). The sharp distinction between the two states seen for the individual in Fig. (**4.2**) is lost when considering a network of interacting individuals. In fact the fraction of elements in one state over the other appears to be a random variable. But that is not unexpected since we have a large number of individuals that are more or less independently making random decisions. The question is what happens when the strength of the interaction is cranked up to the critical value at which a phase transition occurs? It is like watching the weakly interacting particles in a fluid at a temperature above the freezing point and observing the transition to a solid as the temperature is lowered to the critical value, the freezing point, and the force morphs to a long-range interaction that locks the particles into fixed relative positions.

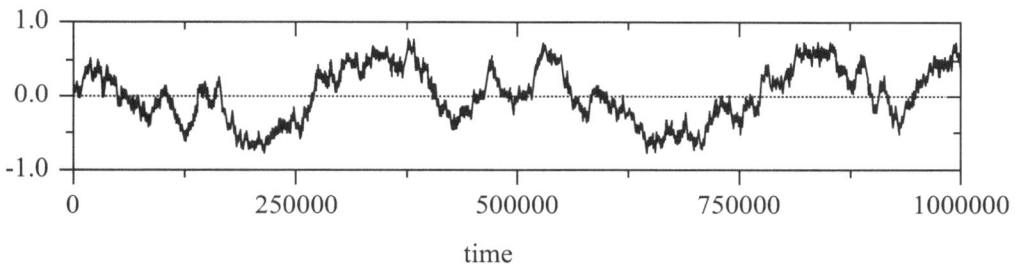

Fig. (4.3). The normalized difference in the number of elements in the network in yes or no state is shown. This variable could also be interpreted as the probability of being in one state or the other. [Adapted from Turalska *et al.* [75]].

One winter evening when I was a graduate student at the University of Rochester I retrieved a bottle of coke from my back porch. I was surprised that the cola had not frozen, since the temperature was well below freezing. As I walked into the kitchen I could see a web of crystals forming within the bottle. The interlacing crystals were the result of an ongoing phase transition that had been triggered by my jostling the soda. The cola slowly turning into ice was the first and last time I witnessed this 'common' phenomenon. It was truly a remarkable thing to watch. So the question is: Does the social decision making model behave in a similar way?

Fixing the interaction strength at the critical value the measure of the global behavior of the network is depicted in the upper panel of Fig. (**4.4**). The network is seen to behave cooperatively, being in either the upper state or the lower state, and it makes very rapid transitions between the two at random intervals of time. One of the interesting properties of the model is that the fluctuations observed in Fig. (**4.4**) persist in the critical state as calculated by Turalska *et al.* [76] for 10,000 coupled elements. In a physical network these fluctuations are a consequence of thermal excitations, that is, the thermal vibrations of the molecules that are locked into a fixed local position. Here the fluctuations are a consequence of the finite size of the network. As the number of elements within the network is made larger and larger, the magnitude of the fluctuations become smaller and smaller; and they do so in a predictable way. Thus, the uncertainty in the behavior of crowds, voting constituencies, audiences and flocks are a consequence of their size. The larger the crowd, the less the variability in the critical state.

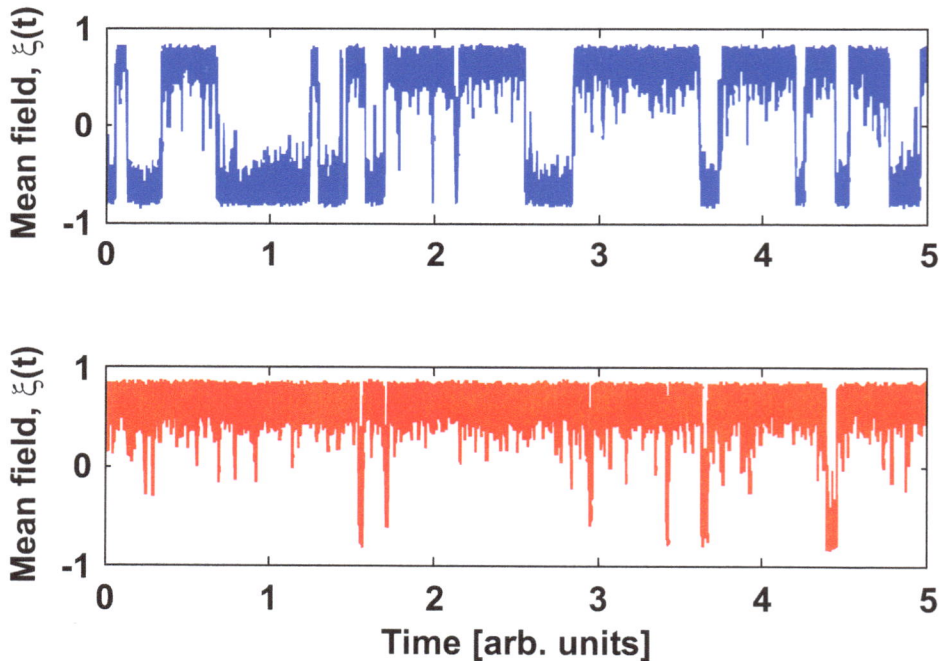

Fig. (4.4). A small number of elements maintaining constant opinion significantly influence the behavior of the network in the organized phase. Top panel: Fluctuations of the global order parameter $\xi(t)$, the fraction of the population excess, for the strength parameter at its critical value and elements located on the vertices of a chess board with a 100 X 100 squares. Bottom panel: The behavior of $\xi(t)$ once 1% of the randomly selected elements are kept constant in time in state "yes" everything else being the same. [Adapted from [76] with permission].

I mentioned that this decision making model suggested a new mechanism to control how a flock of birds appears to change its direction of flight as a unit and how a school of fish, flicker this way and that, but always acting together. The new mechanism has to do with fluctuations [77]. One way to view the fluctuations is as a potential change of state. The network starts to change state, but because there is no bias in the network it quickly snaps back to its original state. A large number of these false starts occur because the conditions are rarely right to facilitate a full transition. The right condition consist of a small bias when the change of state is started that drives it to completion, but this bias is only realized randomly in the absence of an external stimulus. So how does this explain the behavior of a flock of birds in coordinated flight?

When a predator appears it is only seen by birds at the perimeter of the flock or by fish at the edges of the school, and these few elements subsequently act in concert to avoid the predator. They no longer respond to the group, but instead become fixated on avoiding the predator. This uniform action, on the part of a small number of individuals, within the network is sufficient to create the bias that induces the full transition at the next fluctuation [77]. The fluctuations in the cooperative state may therefore be seen as a feedback mechanism that facilitates a collective responsive to changes in the environment. From an evolutionary perspective this flexibility in behavior increases the group's chance for survival.

The abrupt changes in the social organization depicted in Fig. (**4.4**), are not moments of disorder, rather they are instances of increased spatial correlation between the elements of the network. As explained by Turalska *et al.* [76] this condition of extended cooperation, analogous to the critical state of a phase transition, suggests that a few members of a society may exert substantial influence over the entire social network. A similar observation was made in 1975 by Haken [78], who used the concept of phase transition to interpret the 1968 student revolution in France. His model of social interactions explained, in part, the remarkably rapid transition from traditional morality to sexual liberation, however it did so without focusing on the role the minority of protesting students may have played in triggering the change. The concept of a committed minority (inflexible agents), who retain their opinion regardless of their social environment has only recently been introduced [79]. We used the above decision making model to explore Haken's idea to possibly explain the influence that committed minorities exert on both local and global politics.

Consider an inflexible agent to be a randomly selected element at a vertex of the two dimensional chess board that keeps its decision, independently of the opinion of its neighbors. To establish that committed minorities may operate efficiently in spite of their small numbers, in Fig. (**4.4**) we compare the evolution of $\xi(t)$, the average difference in the number of people in the two states, in the absence of a committed minority to the evolution of that average difference in the presence of a small (1%) committed group. From time to time a crisis occurs where $\xi(t) = 0$, when the number in each state is the same. During these times of crisis the network may undergo an abrupt change of opinion and the correlation length becomes large, making it possible for the committed minority to force the social network to adopt its view. As a consequence, during the time interval over which the minority opinion is amplified, it imposes its opinion over the whole network.

Of course this simple decision making model does not fully explain the social phenomena of the Arab Spring or the Occupy Wall Street movement, but like the simple models of population growth in the first chapter it provides a well- defined position from which to initiate a quantitative discussion.

4.1.4. Frequency

Once Mandelbrot introduced the intriguing notion of fractals into science, investigators began to interpret previously mysterious experimental results in terms of this geometric concept. A colleague of Mandelbrot at IBM, Richard Voss, who had made essential contributions to Mandelbrot's books by computationally rendering the fractal art work, which at the time was cutting edge research, had been independently studying a certain kind of fluctuations in physical systems in collaboration with John Clarke [80]. In science a phenomenon is not understood when there is no theory to explain it or, when it has a number of competing theories to explain it. The latter was the case for '1/f noise', a frequently observed phenomenon in which the measured physical variable fluctuates randomly in time and the frequency content of the corresponding power spectrum decreases in inverse proportion to the frequency. The effect had been observed in everything from the variation in the number of electron emitted per unit time from the cathode of a vacuum tube, to the fluctuations in the position of the human eye as gaze is fixed on a given object. It should not be surprising that the fixation of the eye, during a stare, is no more stationary than the heart beat is regular, or that gait size is uniform. All these physiologic phenomena are complex and their time series have inverse power-law spectra, as we discuss in the next chapter.

Remarkably, Voss and Clarke [81] were able to transfer their interests in the 1/f noise of physical processes to social phenomena. They found spectra with the same frequency content in classical music and consequently discovered a relation between music and fractals. Their intent was to quantify aesthetics, or more precisely to quantitatively determine those aspects of music that elicit a positive human response and that might account for why some musical scores become classics and others fade into obscurity. A second purpose of their research was to generate 'stochastic music' on a computer to test their hypothesis regarding the aesthetic appreciation people have for the 1/f spectral content of music.

It is perhaps not surprising that these two scientists should examine how the new mathematical concept of a fractal could be related to music, since it is part of the folklore of science that mathematics and music spring from the same source. It is often pointed out in support of this conjecture that Albert Einstein played the violin, Max Planck was a gifted pianist and Alexander Borodin did research in organic chemistry. It is further conjectured that the way the brain organizes numbers and the sounds we hear are closely related, just as the way we structure geometry through vision is related to the objects we see.

I think that it is probably worth pointing out that part of understanding complex phenomena is to reduce what is not understood to one or a few basics that are only a partial mystery. In this way it is not the many daily events people encounter whose understanding eludes them, but a single underlying cause that gives each event its separate shape and form. These disguised manifestations may be pleasing to behold or not, but they are all different representations of one underlying incompletely understood truth. Herein I trace most if not all of what people do not understand to complexity, which is manifest through 1/f variability. I do not address these lofty goals for the moment and instead return to the music.

Voss and Clarke determined that music has a blend of regularity and spontaneity that is characteristic of 1/f variability and the time series corresponding to such spectra are intermittent. Fig. (**4.5**) shows the measured power for the frequency fluctuations necessary to physically drive the speakers for several different musical compositions. Each musical piece has the same qualitative behavior of decreasing power from low frequency to high frequency.

On the other hand, white noise has the random signal depicted on the top left in Fig. (**4.6**) and has all frequencies of equal amplitude within the signal, as indicated by the horizontal line for the spectrum. Such noise was given its name by Wiener, who was studying the properties of white light at the time and noted that white light consists of the incoherent superposition of all visible frequencies with equal energy content. Such a noise spectrum is indicative of a process with no memory of its past; what happens at one instant of time is completely independent of what happens at any later instant of time. Brown noise, depicted on the bottom left in that figure, has frequencies that mimic the erratic movement of a Brownian particle in a fluid. The spectrum can be seen to nose dive to zero at high frequency inversely as the square of the frequency. Such phenomena are strongly dependent on their immediate past.

The empirical 1/f spectrum of music shown in Fig. (**4.5**) falls midway between that of computer generated uncorrelated white music, with its flat spectrum, and computer generated brown music, with its steep quadratically descending spectrum. Consequently 1/f noise is also known as pink noise falling as it does between white and brown noise. Voss and Clarke [82] offered for evaluation the three kinds of stochastic music to a number of people from a variety of backgrounds. The overwhelming majority selected the pink music as more aesthetically pleasing than either brown or white music. This choice suggests that scaling captures one of the essential features of music considered pleasing on a historical time scale. For example, the music of Bach, Davidovsky, and Jolas, share the ubiquitous 1/f behavior as shown in Fig. (**4.5**).

It is not only about regularity and variability in music that humans make aesthetic judgments. They do the same when examining photographs and paintings. Musha [83] reported on a study of the distribution of the breakup of space, which is to say the spatial composition, in a number of paintings. In the three decades since his work was reported, a number of other scientists have scanned paintings and photographs, using levels of contrast as a measure to record the frequency of occurrence of spatial scales. The 1/f phenomena in spatial scale was observed again and again in classical art work. This is particularly true in the cubist work of Picasso. The subsequent evolution away from representational art has lost this contact with 1/f variability and a subsequent loss of appeal of 'modern' art on the part of the general public.

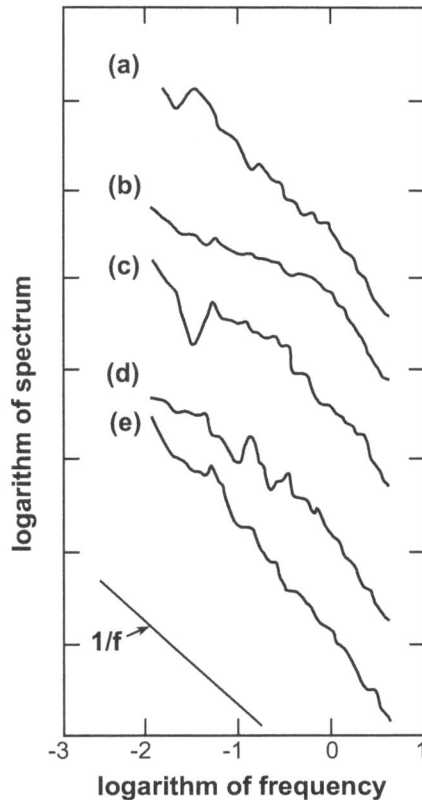

Fig. (4.5). Audio power fluctuation spectral densities versus the dimensionless frequency for (a) Davidovsky's Sychronism I, II, and III; (b) Babbit's String Quartet number 3; (c) Jolas' Quartet number 3; (d) Carter's Piano concerto in two movements; and (e) Stockhausen's Momente. [Adapted from [82] with permission].

It appears that pink noise has more than just aesthetic appeal. It was also reported [83] that 1/f variability is evident in other aspects of biological networks; for example, it plays a role in the relief of chronic intractable pain for which transcutaneous electrical nerve stimulation, with a 1/f spectrum, is beneficial.

One of the oldest recorded medical procedures is acupuncture dating back before recorded history in China as suggested by the archeological finding of stone needles. In the West the image of acupuncture is that of treating patients by inserting and manipulating needles in various regions of the body. But it is also part of the eight thousand year Taoist tradition in which people meditate to observe and control the flow of energy from the universe, through the relation between the human body and nature. Jesuit missionaries brought reports of acupuncture to the West in the sixteenth century, and it was adopted at that time in France. However acupuncture has never been universally accepted by western medicine, being adopted by some countries at different times in history, only to be subsequently rejected by their medical communities.

Recently, traditional acupuncture has joined with modern technology to replace the manipulation of the needles with transcutaneous electrical nerve stimulation [83]. In various treatments the response to different kinds of random electrical simulation were compared. An electrical signal was applied in these experiments to disrupt pain messages to the brain, and the frequency content of the signal determined the success of the interference. White noise stimulation has the random signal depicted on the top left in Fig. (**4.6**). The response of patients to white noise was reported to be too erratic and sometimes elicited a fear reaction. Brown stimulation has the random signal depicted on the bottom left in that figure and was reported to be monotonous and displeased patients. The reported response in the majority of cases was that a signal with frequency content mid-way between white and brown noise was preferred. This pink signal successfully reduced pain levels associated with back pain, trauma, cancer, and lumbago for significant periods of time after stimulation was removed.

Here we have described two contrasting ways in which 1/f variability enter our lives. The first is through the aesthetic judgments we make regarding music and art; the second is the perception we have of pain in our bodies. My friend Bill

Deering conjectured with me [84] that the same pink signal, that can soothe the savage beast, can also relieve the beast of his pain. We speculated that in a way we did not understand at that time, a pain-free life and a joyfully aesthetic one, are related in some biologically fundamental way. We remarked that such speculation must await the carefully devised experimental test. That test has now been done.

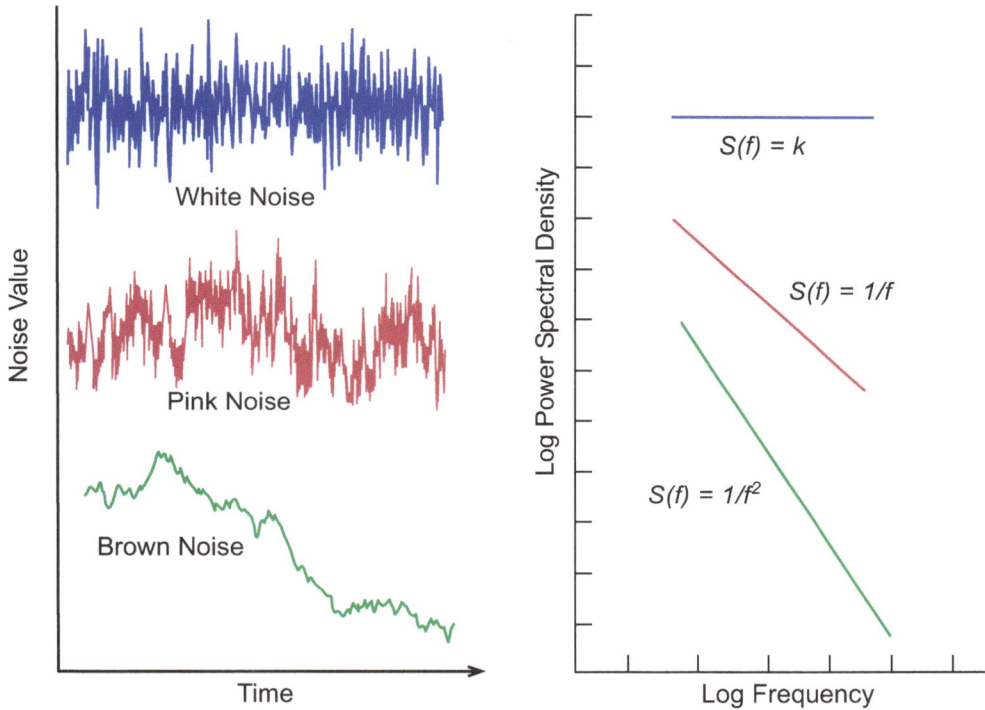

Fig. (4.6). Typical time series are shown on the left for the various kinds of noise. The top left depicts a white noise time series having the flat spectrum of frequencies on the top right. The middle left is a pink noise time series with less variability than white noise and a 1/f spectrum depicted on the middle right. The lower left Brown noise time series has the least variability with the 1/f² spectrum depicted on the lower right.

Scientists have determined that 1/f signals are encoded and transmitted by sensory neurons more efficiently than are white noise signals [85]. These experimental findings are consistent with recent models that explain how and why some information is efficiently transferred between complex networks and other information is not [71].

4.2. SCALING MECHANISMS

I almost never read a review before going to a movie, which a British friend finds scandalous. I go to a movie when I like one or more of the actors, or when it is an adaptation of an author whose work I admire, or people I respect won't stop talking about it. The same is true for attending a seminar at the university; I go when I know and respect the research of the lecturer, or at a colleague's urging. I suppose you can call this a social attraction mechanism. If an individual has this star power, they usually draw a large audience for their movies or lectures. This particular mechanism by which social networks are formed, has been drawing a great deal of attention recently, but it is hardly the only one. Scientists interested in understanding inverse power-law phenomena have identified a large number of distinct mechanisms that purport to explain how complex networks dynamically assemble themselves over time. Andriani and McKelvey [86] have identified 15 different mechanisms, not all of which have subsequently been found to be distinct, which lead to scaling. The mechanisms produce scale-free networks; networks whose statistics are described by probability distributions not characterized by one or a few scales.

Probabilities are not the only mathematical descriptors that scale, another is that of time series. Fig. (**4.7**) shows a time series consisting of the beats per minute of a healthy human heart. The variation in the heart rate over time for three hundred minutes is depicted by the top curve in the figure. The degree of variability is surprising and stands in marked contrast to the phrase 'normal sinus rhythm' heard in the conversations of physicians. The time series certainly does not have a strong rhythmic component. The part of the upper curve in the box is magnified in the middle graph showing the heart rate variation over thirty minutes. The time scale has been decreased by a factor of ten in going from one curve to the other. It is curious that the top and middle curves are so statistically similar, given that their time scales are so different. The bottom curve contains the information content of the boxed part of the middle curve for another reduction of a factor of ten in time scale. Again the bottom curve does not look too different from the top curve and yet the time scales in the two graphs differ by a factor of one hundred. This statistical resemblance implies that the distribution of fluctuations in the time series scale, which is another way to say that the fluctuations in time are fractal or self-similar.

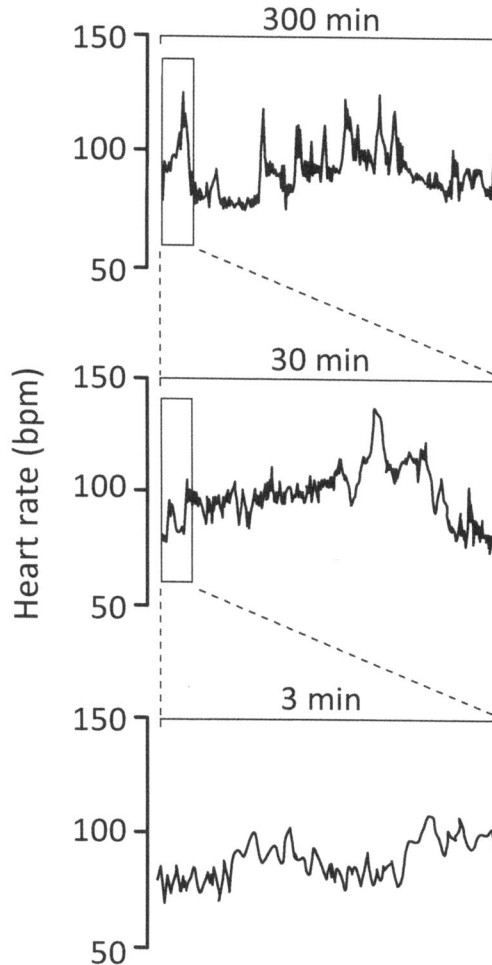

Fig. (4.7). The heart rate data for a healthy human heat is depicted for three different time resolutions. The figure depicts the scale-free nature of the heart rate time series.

Heart rate variability is used here to represent a large class of complex phenomena in which the time series for the variable of interest scales. The fluctuating process appears pretty much the same at every resolution scale on which it is viewed. This statistical self-similarity is modern scaling and the similarity in the variability at each scale does not occur by accident. There must be one or more mechanisms that control the degree of variability at each level of resolution. One mechanism to control variability across scales is through correlations, so that what happens at a given time is influenced by what happened at adjacent times. Such influence is determined by a power-law index in the variance of the time series, which is to say that the variance increases in direct proportion to a non-integer power of the time. This power-law or scaling index is one possible measure of the complexity of time series and to understand the source of complexity we need to identify mechanisms that can generate such scaled time series.

As mentioned, the scaling of time series is not the only way a complex phenomenon can scale. If the probability density scales in time, then so does the second moment. This behavior was alluded to in the discussion of the diffusion of oil slicks. The statistics in simple diffusion are Normal, but the variance increases linearly in time. More complex physical processes, known as anomalous diffusion, can have Normal statistics, but in addition they have a variance that grows algebraically, but not linearly in time. In the case of anomalous diffusion, the power-law index is not one, but has a value characteristic of the underlying physical process that induces correlation and that index can have a wide range of values.

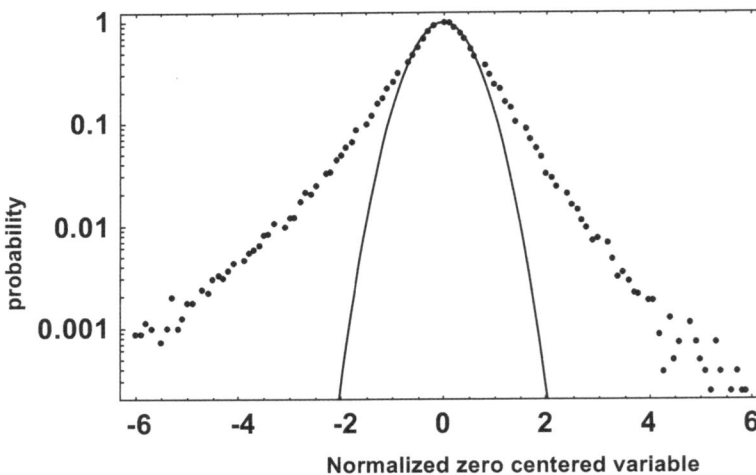

Fig. (4.8). The empirical relative frequency of the differences in the heart's beat-to-beat intervals. To facilitate comparison the variable is divided by the standard deviation of the increment data and the relative frequency is divided by a constant. The solid curve is the best Normal distribution fit to the data and grossly underestimates the tails of the distribution. [Adapted from [46] with permission].

But it gets even more complicated. Suppose the complex phenomenon scales, so that its second moment increase as a power law in time, but its statistics are not Normal. The preceding discussion might imply Normal statistics of the heart rate time series, when in fact, it is not. These two facts taken together, the scaling of the second moment and the non-Normal statistics, implies that the statistical distribution of the heart rate time series itself scales. Fig. (**4.8**) shows the statistical distribution determined, using twenty-four hours of heart beat data from a single individual. The data points consist of the differences between consecutive times shown in Fig. (**4.7**), with the average value subtracted so the data are zero centered and the original data are divided by the standard deviation of the beat interval data [46]. The solid curve is the best Normal distribution fit to these data. It is clear from the figure that the actual distribution has very long tails; tails that, in fact, satisfy an inverse power-law. We therefore need to identify mechanisms that can generate a scaled inverse power-law probability.

It is time to bite the bullet and describe some of the many mechanisms that may account for the observed scaling behavior in time series and the inverse power-law probability densities observed in complex phenomena. In the following nine subsections we sketch in no particular order what might be considered the most significant mechanisms that have been shown to generate scaling and Pareto's inverse power-law distributions.

4.2.1. Hierarchical Effect

The notion that the activity observed in natural phenomena can be a consequence of many invisible layers of movement and that this imperceptible layering relates successive levels is the hierarchal effect. In 1733 Jonathan Swift in reflecting on self-similarity considered reduced versions of what is observed on the largest scale, being repeated in an ever decreasing cascade of activity to ever smaller scales, wrote:

> "So, Nat'ralists observe, a Flea;
> Hath smaller Fleas that on him prey,
> And these have smaller Fleas to bit'em;
> And so proceed ad infinitum."

Regarding this effect, Mandelbrot observed that when the process is truly self-similar, a fractal is formed. However fractals are not only formed by scaling downward, they also blossom upwards as anticipated by de Morgan in 1872:

"Great fleas have little fleas upon their backs to bite'em;
And little fleas have lesser fleas, and so ad infinitum;
And the great fleas themselves, in turn, have greater fleas to go on,
While these again have greater still, and greater still, and so on."

Poets, no matter how lacking in skill, capture an essential feature of nature in which a complex network forms a self-similar hierarchy. A network is formed by linking subnetworks, and each subnetwork in turn is formed by the linking of its own subnetworks, telescoping towards ever smaller scales. Simon [87] argued that hierarchic systems, or hierarchy, is one of the central structural schemes found in the architecture of complexity. He emphasized that this use of the word hierarchy does not imply that the subnetworks are subordinate to the network they form. There is no 'boss' in the hierarchy, or at least such an authoritative figure is not implied. He distinguishes between hierarchy and 'formal hierarchy' that is used in the more traditional sense of governments, militaries and universities, which clearly have an authoritarian top-down structure.

A hierarchy can be seen as a multi-level structure. At each level the elements are complex and linked, and at the next level down each complex element is made up of a number of different complex elements with even stronger linkages. An organ is made up of connected tissue; the tissue is composed of compacted cells; the cells in turn consist of nuclei, cell membrane, microsomes and mitochondria. At each level of the hierarchy we define a new kind of force to describe how the elements are linked together. Neutrons and protons share a nuclear force to hold together the nuclei of atoms; the atoms share a Coulomb electrostatic force to stabilize molecules; the molecules in turn produce a van der Wall's force to hold together solid matter; aggregated matter produces the gravitational force that stabilized planetary systems and galaxies.

The various force laws distinguish the different kinds of hierarchies that can be formed. In self-similar hierarchies all the elements at a given level are linked to a given number of elements in the previous level. All links have the same constant strength and their number increases as a power of the level number. The probability of forming a connection between elements as the number of levels becomes very large decreases as an inverse power law in the number of connections. The power-law index depends both on the number of new elements appearing at each new level and the linking of these elements. This forms what is called a hierarchal web [88].

4.2.2. Entrepreneurial Effect

How do we make the transition from the eminently fair Normal distribution of income that is not observed, to the demonstrably unfair, but universally observed inverse power-law distribution of income of Pareto? One way to understand how such a transformation between distributions can occur is by shifting attention from independent individual incomes to include the effect of collections of individuals joining together for a common purpose, for example to form a company. Technically this is one way to overcome the independence requirement necessary for the central limit theorem. An initial distribution with a given finite variance can incorporate an additional component with a broader width due to the additional people involved in forming a small company and this occurs with some small probability. Of course these small companies may in turn join together to form a larger company so that the modified distribution can be further modified to include an even broader width due to the aggregating of small companies and this occurs with an even smaller probability. In this way one can construct a geometric series for the probability distribution, each term of the series being given by the initial Normal distribution with a geometrically increasing variance to incorporate the influence of more and more groups of people and each increase occurring with a geometrically decreasing probability.

Take the initial distribution in income to be exponential, since the form of this distribution is irrelevant to the argument. It is clear that the broader distribution, the one that includes the amplification factors through the formation of companies, places more and more of the variance out into the tail. The amplitude of each of the additional terms is small because the probability is increasingly lower, but as the series of amplification terms becomes longer the contribution to the tail builds up. This build up is depicted in Fig. (**4.9**), where the initial distribution was taken to be exponential. We see the formation of an inverse power-law tail in the distribution as an infinite series of the amplification mechanism modifies the distribution.

This amplification mechanism is at the heart of entrepreneurial job growth in an economy and was first developed in this context by Montroll and Shlesinger [89]. Their theory actually relates the power-law index to the ratio of the size of the amplification factor and the probability that a factor of that size occurs. Using economic data from the United States they predict an inverse power law of the right size emerging, a Pareto index of approximately 1.5 through the formation of companies having between 15 and 20 people. Thus, the imbalance implicit in the Pareto distribution can be triggered in an economic context by a mechanism that many economists feel is necessary for capitalism to prosper, that being the entrepreneurial effect. Of course this mechanism only explains the tail of the income distribution and more detailed arguments are necessary to explain the existence of a middle class. However, the tail may be necessary to insure the persistence of the middle class over time.

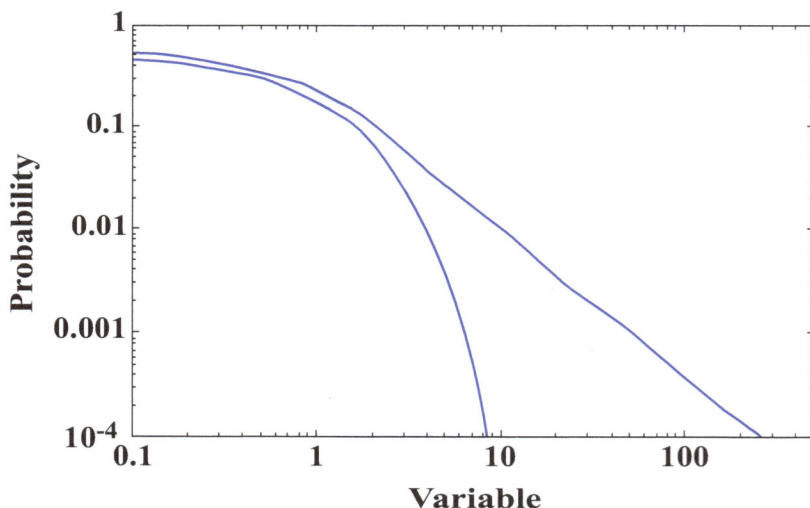

Fig. (4.9). The initial hypothetical distribution of income is exponential and is given by the lower curve. The upper curve is the superposition of an infinite number of such hypothetical exponential amplification terms that give rise to an inverse power-law distribution asymptotically, that is, the observed Pareto distribution of income.

The middle class is modeled by the initial distribution in the absence of the Pareto tail. One such distribution that has a great deal of economic data support is the log-normal and a detailed discussion of the stability of a model economy and the Pareto imbalance is given by Nicola Scafetta and myself [90]. So you might wonder why I choose to use the exponential in Fig. (**4.9**) rather than the log-normal. There are two reasons. First is the fact that any initial distribution with a finite variance does not affect the formation of the power-law tail and I wanted to emphasize that point. Second is that the entrepreneurial mechanism under different names has been used to generate the Pareto tail in a number of contexts. One area of particular interest is that of cognition. The phenomenon of adaptation has been modeled using multiple exponential processes to account for the observed range of time constants in neuronal networks. Just as the economy uses this mechanism to leverage investment, the brain may well use it to leverage how we think.

4.2.3. Allometry Effect

Allometry is the study of the influence of the size of an organism on the function an organ performs and dates back two hundred years to the experimental observation that the size of the brain increases with the size of the animal. Allometry has a particularly simple mathematical form expressing the function of interest, such as an average metabolic rate, the average rate at which an organism generates heat, to an organism's size raised to a non-integer power. The size of an organism is typically expressed in terms of the average total body mass and the resulting equality is called an allometry relation. The proportionality constant in an allometry relation is the allometry coefficient and the power-law index is the allometry exponent. An example of such an allometry relation across multiple species was given in Fig. (**2.9**) for species spanning weight scales from that of a mouse to that of an elephant, a range of ten thousand to one, and the allometry exponent has the empirical value 0.73.

Many theories have been constructed to explain allometry relations and the values of the allometry parameters. The first theory of allometry was devised to identify the reason why small animals generate heat faster per unit mass than do large animals. The phrase per unit mass of a quantity means that the quantity is divided by the total body mass. The original argument to determine the allometry exponent in a metabolic allometry relation was geometrical and based on the idea that the heat loss from an organism is roughly proportional to its free surface [91, 92]. The heat generated by an organism is proportional to its total body mass since the number of chemical reactions generating the heat is proportional to the volume. The balance between heat generation and heat loss is achieved through a matching of the metabolic rate (heat generation proportional to volume) with the animal's free surface area (evaporative cooling proportional to surface area). Consequently, the metabolic rate is proportional to total body mass raised to the 2/3 power; the 2 from the surface area and the 3 from the volume. This is the theoretical value determined by Sarrus and Rameaux using this geometric reasoning in 1839 [92].

The experimental 'proof' of the 2/3 value for the allometry coefficient in the 'surface law' was made by Rubner [93] in 1883 over half a century after the geometric argument was formulated. Rubner was the first to experimentally examine the relationship between heat production and surface area [91]. He determined that the ratio of the heat produced by dogs to their surface area was nearly constant, thereby supplying experimental confirmation of the surface law. Of course, as Calder [91] pointed out, despite its allure the 'surface law' is not a law in the sense that gravitational force is a law, but is more properly described as a 'surface rule'.

The expectation was that further observations would support the reasoning that the metabolic rate would be proportional to the 2/3 power of the total body mass as prescribed by the surface rule. Surprisingly, this expectation was not met. The research of Kleiber [94] revealed that the empirical slope in the log-log plots of metabolic rate *versus* total body mass were closer to 3/4 than to 2/3. Subsequent observational studies have reinforced this allometric pattern observed in the data, including some relating the 3/4-rule to plants [95]. Due to the fact that some data sets yield 2/3, whereas other yield 3/4, the empirical value of the allometry exponent remains controversial [96].

The mechanism identified here as producing the allometry relation is one of biological stability, according to the theoretical reasoning. If the balance between heat generation and cooling is not maintained the organism becomes unstable; it overheats and unless this temperature rise is brought under control the organism dies. Consequently, small animals must generate more heat per unit mass than large animals to compensate for the heat loss from their larger surface area per unit mass.

The geometric explanation of the allometry relation is fundamentally a balancing between two competitive influences to maintain stasis. When nature evolves a mechanism that works scientists subsequently find different forms of that mechanism in a wide variety of phenomena. The kind of balancing observed here to explain allometry relations is no exception. It appears again in biological speciation, that is, in the number of new species occurring in time and over space. The naturalist Taylor [97] investigated questions such as determining the variety of species of beetle that can be found in a given plot of land. He determined that by sectioning a parcel of land into grids of ever smaller sizes and sampling the number of species in each square of each grid he was able to construct an empirical relation between the variance and mean of the number of new species observed. The allometry relation he constructed had the variance in the role of function and the mean playing the role of size; he did not identify his rule as an allometry relation.

Subsequently, Taylor and Taylor [98] postulated a behavioral model to account for the speciation allometry relation. In their model individuals act under conflicting ecological pressures. On the one hand individuals move under selection pressure to maximize their resource; they spread out. On the other hand individuals move to maximize environmental quality and seek out company so as to efficiently reproduce. Each of these pressures has a different fractional power of the population density and the struggle between the attraction of one mechanism and the repulsion of the other is universal. Consequently it is the dynamic balancing of the two, migration and congregation, that produces the interdependence of the spatial variance and mean population density.

Another example of the operation of the allometry effect is in the operation of the human heart and the resulting variability of the interbeat intervals. As Goldberger and West [99] briefly review, the sinus node (the heart's natural pacemaker) receives signals from the autonomic (involuntary) portion of the nervous system, which has two major branches: the parasympathetic, whose stimulation decreases the firing of the sinus node, and the sympathetic, whose stimulation increases the firing rate of the sinus node pacemaker cell. The influence of these two branches produces a continual tug-of-war on the pacemaker firing, resulting in fluctuations in the heart rate of healthy subjects. The heart rate time series scales and the scaling index is a result of this competition mechanism. Allometry relations model the opposing ecological pressures of biological speciation in Taylor's Law and the heating and cooling balance of the surface rule.

4.2.4. Gambler's Ruin Effect

At sixteen, when I had money, I would enjoy the company of some shady characters in a local pool hall. The pool hall was one big room on the second floor of a turn of the century building. There were six or eight pool tables with overhanging lights that were turned off unless there was action on the table. Just after dinner the pool hall would be dark, with the only light being over the counter just to the right of the door at the top of the stairs. Sitting on a stool behind the counter there was a well-dressed skinny fellow, wearing a Fedora, with a wandering right eye who would clock time at the tables, turn the lights on and off and collect money. He was also a gambler.

One of the gambling games he enjoyed and which two could play was coin matching. The game was simple. One player would call odd or even and each would shake up a number of coins and quickly place them in a row along one arm. The arms of the two players were then lined up and the coins compared. The player who called even would take the coins that matched, leaving the mismatched for the odd. It was possible to run through quite a bit of money that way in a very short time. But the game was fair and I had an equal chance of losing everything, or making a good deal of money (at least for a 16 year old), or so I thought. I was not then aware of the gambler's ruin effect.

The matching game was fair, because each comparison of coins had a fifty-fifty chance of being the same or different, therefore the probability of being odd was 1/2 and of being even was also 1/2. The game was therefore formally the same as the drunkard's walk with each matching determining if the walker steps to the right or the left. Adding all the comparisons together determines how much a player won or lost, or how far the drunk was from where she had started. This is clearly an unbiased random walk process. The game ends when either player decides to stop, or the more usual situation in which one of the players loses their bankroll; the gambler is ruined. The probability of achieving the number of matches required to reach this level of exchange defines the gambler's ruin problem.

This problem can be cast in a form we have already seen. Suppose each match takes the same amount of time, then we can describe the gambling process in terms of time taken rather than in terms of the number of matches. Further we can identify the bankroll, in this case the amount of money I had, which was very much less than what the proprietor had, with an extrema. When the walk intercepts the extreme the game ends. The gambler's ruin problem is then equivalent to determining the distribution of times required to reach the extrema for the first time. The point is that if one person has a significantly larger bankroll than the other it is just a matter of time before the under- financed player is wiped out. Going broke does not have the level of significance of the power outage we encountered earlier, but its emotional impact on a 16 year old is probably more immediate.

The gambler's ruin problem can also be represented by the difference between the size of the bankroll and the amount of money lost. The gambler's ruin then occurs when the trajectory of the game first intercepts zero, at which point the player has lost everything. If n is the number of steps (time or comparisons) then the probability of intercepting zero for the first time is determined to be a Pareto distribution in the number of steps with an exponent of -3/2. Therefore unless I quit early in the evening I was destined to lose everything I had; it was just a matter of time, particularly because the distribution of extrema is inverse power law. The likelihood of going broke in a relatively short time was quite high. Of course what I remember are the times I won and not those I lost, partly because my friends usually arrived before I went broke and we would shoot pool. There was less luck involved in pool than in matching coins.

4.2.5. Matthew Effect

Scientists have a communication network that has been cultivated and refined over the past three centuries. There are multiple modes of communication between scientists: publishing papers and books; exchanges in private letters and individual conversation; talks at conferences; seminars and colloquia at universities; and now there are emails and the Internet. Individual scientists acknowledge the influence of the research of others on their own work, through the process of citation, where they explicitly mention the publications they have read, talks they have heard, and describe how that work is related to their own investigations. Such citation can significantly impact scientific careers for better or worse. Recall that Einstein mentioned the work of the botanist Brown in his discussion of diffusion, although he was initially not sure exactly how it related to his own work, and was not aware of the published work of Bachelier on the same phenomenon. Consequently today the community refers to Brownian motion and Einstein diffusion and most scientists have not heard of Bachelier and his theory of diffusion.

In this referencing process more credit is typically given to the senior authors of papers than to their more junior colleagues, who might be students or post doctoral researchers. This is not done so much in the publication of papers in the physical sciences, where the junior member is often afforded the privilege of first author in the paper's byline. However, it is all too often evident in informal conversations, where such imbalance in recognition is clearly revealed. Of course there is injustice in such attribution as was extensively and eloquently discussed by the Nobel Prize winning sociologist Robert Merton [100]. He referred to this as the Matthew Effect in science and recounted a theme running through interviews with Nobel laureates [101]:

> "They repeatedly observe that eminent scientists get disproportionately great credit for their contributions to science while relatively unknown scientists tend to get disproportionately little credit for comparable contributions. "The world is peculiar in this matter of how it gives credit. It tends to give the credit to (already) famous people."

This effect is not limited to apportioning credit in science, but can be found in essentially every element of human activity. To paraphrase Merton the Matthew effect consists of recognition being given to the already famous, to the detriment of those who have not yet made their mark. This effect is misallocation of credit and its name stems from the Gospel according to St. Matthew:

> "For unto every one that hath shall be given,
> and he shall have abundance:
> but from that hath not shall be taken away
> even that which he hath."

Merton goes on to point out that this effect may act to enhance the visibility of new scientific communications, by having the name of a famous scientist on a publication. In this way the effect facilitates communication of new results and science is nothing if not a social activity. Scientific findings are of little or no value until they are shared, analyzed, tested and finally accepted by the appropriate segment of the scientific community and subsequently become part of the body of human knowledge. Therefore, although it may be dysfunctional for certain individuals within a social network, the Matthew effect can be functional for certain general aspects of the same social system.

In addition to its reward and communication aspects the Matthew effect has also been expressed as the principle of cumulative advantage, identified by de Sola Price [20]: "The rich get richer and the poor get poorer". This principle identifies a mechanism of attraction by which the most prestigious universities attract the best students, and which Merton substantiates by identifying the top six universities at that time, 1968, and noting that they produced 23% of the doctorates in the physical and biological sciences. Of course, the situation has changed in the last half century but the principle remains valid.

In his excellent book [102] Watts explains in terms of the Matthew effect, how complex networks such as the Internet grow over time. As new nodes enter a network they are preferentially attracted to those nodes with the greater number of links. As they link up with those with the greater number, they increase the imbalance in the number of connections to each node. Those that have been in the network the longest, by this argument, have a statistical advantage in terms of the number of new links they attract. Consequently the older nodes in the network have an advantage over the younger. This mechanism was introduced as *preferential attachment* by Barabási and Albert [103] into the study of the growth of complex networks to explain the inverse power-law in the number of links connected to a given node. This mechanism was, in fact, identified nearly a century earlier by the mathematician Yule [104] in his collaborative investigation into the biological mechanism responsible for the generation of new biological species.

The first scientific investigation to motivate the construction of an inverse power-law model was that of the evolutionary biologist Willis. In 1922 Willis hypothesized that the physical area occupied by a biological species is directly proportional to the species age, so that area becomes a measure of age and he collected a great deal of data in support of his thesis [105]. Yule [104] explained Willis' observation that the number of genera is described by an inverse power-law by introducing a model in which the size of the biological genera is determined by the number of mutations occurring over the time scale of biological evolution.

4.2.6. Modulated Fluctuations Effect

A different kind of mechanism leading to inverse power-law behavior is analogous to the thermodynamic balance between molecular agitation and the macroscopic movement of material. Recall the phenomenon of diffusion where the hot particles of the coffee disrupt the motion of the heavier milk particles in the cup. The random agitation of the milk eventually produces a uniform tan color in the milk-coffee mixture. In his investigations of Brownian motion, Einstein established that the ratio of the strength of the fluctuations to the dissipation rate is proportional to the temperature of the coffee. This finding was so significant that it was given a name. It was the first *fluctuation-dissipation relation* and established that the rate at which the influence of perturbations fade and the strength of the statistical fluctuations in the system are not independent of one another, but are tied together by the physical temperature of the environment.

At least two forces are in balance during diffusion, the random force of the particles of hot coffee, acting on the milk and the linear dissipation quenching the movement of the milk particles. The dissipation is the result of drag on the heavier milk particles produced by their pushing the lighter particles of the viscous coffee out of the way as they move. Drag is a macroscopic effect and is also observed in the motion of ships as they plow through the water on the ocean surface, the rougher the surface the greater the drag coefficient or dissipation. The drag is proportional to the velocity of the vessel, or in the case of diffusion, to the velocity of the milk particle. This balance of linear dissipation against linear additive fluctuations produces Normal statistics in the velocity of the milk, as discovered in classical diffusion. However, the situation is quite different when the random fluctuations themselves depend on the milk particle's velocity and therefore the process can no longer be considered linear and additive.

The loss of the linear additive nature of the dynamics is the starting point for constructing many mathematical models with properties different from those of the physical process of diffusion. The modification of the diffusion paradigm of concern here has to do with the random force. Suppose the random force is proportional to the velocity of the milk particle in which case a high velocity amplifies the random force and a low velocity suppresses the random force. This modulation of the random fluctuations is sufficient to violate the fluctuation-dissipation relation and therefore results in non-Normal statistics. If we retain the linear dissipation, the resulting imbalance produced by the modulation of the fluctuations, give rise to inverse power-law statistics. The power-law index in the inverse power-law distribution is determined by the ratio of the strength of the perturbation to the rate of dissipation [106].

This argument suggests that for a thermodynamically closed system with a linear additive random force, such as occurs in diffusion the statistics are Normal. On the other hand for a thermodynamically open system with a linearly modulated random force the statistics are inverse power law. In the closed system the linear dissipation is sufficient to balance the energy input by the random force, but in the open system it is not. In fact one can think about the latter system as being produced by a fluctuating dissipation rate, the average rate is the linear dissipative part of the force and the fluctuating part of the rate is the linear modulation of the random force.

Consider a boat floating adrift on the sea surface. When the surface wave field or sea state is low there is a linear drag on the boat and its motion on the surface is Normal, as determined from the behavior of the oil slick. However in a high sea state the random water wave field induces a fluctuating component to the drag and the boat's movement is no longer Normal, but inverse power law. Of course in this latter situation the imbalance can produce instabilities and the freely responding boat may well flounder.

4.2.7. Contagion Effect

The contagion effect has changed significantly from its 14th Century English roots defining it as having contact with, or polluting. In a medical context contagion refers to the spread of disease and the growth of epidemics. More generally the effect encompasses the infectious nature of laughter, where a laughing child infects an onlooker causing an adult to laugh in unison. A skilled comedian uses this effect to steamroller the first few laughs of an audience into a torrent of guffaws. The effect also describes how rumors spread within a group and a population becomes no less infected with an idea than with a virulent disease. The contagion effect is also known in finance as the domino effect, where a default by one institution can generate a cascade of failures within a financial system. Recall the geopolitical domino theory of the spread of communism from the sixties. The equivalence of the terms contagion effect and domino effect is however limited to a linear process where no member of the population can be immune.

The previous discussion of population growth models focused on biological populations, but there was nothing intrinsic to the models that excluded considering such phenomena as rumors, epidemics or ideas. These latter populations do however require mechanisms for growth that differ from those of Malthus and Verhulst. It is no longer the fecundity of the population that determines growth, nor is it the sanitary conditions that determines the saturation to that growth. In a social context the number of links between population members determines the rate of growth of the number who have heard the rumor. For the spread of disease assume a person coughs within a group and for each cough the same fraction of the group incubates the virus and comes down with a cold. If this constant transition mechanism operates for each person then the growth in the number of those with a cold is exponential. However, like the growth of biological populations, such increase cannot last forever and it eventually saturates.

The influence of one population on another occurs here with the interactions of the populations of susceptibles, infectives and recovered, which can be reduced to the interaction between susceptibles and infectives. What is often of most interest is the growth of the infective population. Vlad and Schönfisch [107] identified two universal scaling laws, corresponding to intermittent and non- intermittent fluctuations. If the fluctuations in the number of infectives is not intermittent then the probability of becoming infected is exponential. On the other hand if the fluctuations in the number of infectives is intermittent, that is, the fluctuations occur in bursts over time, the probability of becoming infected is inverse power law. They prove that these results are universally valid for high migration systems, that is, for networks in which the elements move around freely, as the size of the system and the size of the population becomes very large with the population density, the ratio of the two, remaining finite. Under these conditions the universal laws arise independently of the form of the interaction (linkage) between the two populations.

4.2.8. Criticality Effect

As its name implies, the criticality effect is the influence a critical (tipping) point in the dynamics has on a complex network. Avalanches and collapsing bridges are the result of networks exceeding their critical point. But the effect of a critical point is not always so dramatic. The mundane freezing of water and the evaporation of perspiration from skin are also examples of phase transitions, where a system parameter (temperature) achieves a critical value. At the heart of the discussion of the transitioning of gases to liquids, and liquids to solids, is the notion that as a physical parameter is changed, in this case the temperature, the system passes through a critical state in which the correlation length, the distance over which elements influence one another on average, becomes extended and can even become infinite. In a social context the onset of such long-range effects are equally obvious. The severity of the crime, or its media visibility, can extend the manhunt from the town Sheriff to the FBI.

In a physical critical state the interaction between particles expands its reach from local to global. As the temperature is lowered and passes through the critical temperature for liquid/solid transition the local coupling becomes non-local and loosely coupled particles in liquids transition to tightly coupled particles in solids; water becomes ice. In a social context the apparently random motion of pedestrians in a city park under the proper conditions can become a synchronized army. The correlation length and correlation times in a critical state tend to diverge and the well behaved functions that described the smooth dynamics are replaced by inverse power laws with critical exponents.

The criticality effect has been observed in social phenomena, where individuals are asked to make a choice, say between two candidates in an election. Prior to the election season the choice could well be random, with an individual's vote changing over very short times, say on a daily basis. Decision making models [45] have been developed that parameterize the coupling among people in the electorate and as the strength of that coupling increases a critical value is reached. The fluctuations in the consensus is determined by the number of people in the network (there is no temperature in the social situation) and the correlation length increases in the critical state to eventually match the size of the population. Prior to the critical transition each individual has an independent opinion either for or against a given candidate and the correlation of their vote with that of their neighbors' vote is essentially zero. Fig. (**4.3**) shows the corresponding time series fluctuating above and below zero. At the critical point the dynamics of the network change from local to global and the fluctuations entrain more and more people thereby extending the reach of the correlation as depicted in Fig. (**4.3**). Consequently, once the transition is complete the voters no longer act independently, but instead they act as a constituency, either all for, or all against, a given candidate.

Of course the models of voting and decision making are not predictive in the sense that they do not faithfully predict who will be the winner of an election. Such models at best determine the probability of the electorate shifting its collective vote from one candidate to the other. In Fig. (**4.3**) the consensus is seen to be for the upper candidate, but there are short-time fluctuations in the number of voters due to the finite number of individuals in the electorate (model). In this particular model of decision making it is possible for consensus to shift completely from one candidate to another, making it perhaps more suitable for modeling a primary than an election. It would not be appropriate for modeling an election between two parties since the majority will not change political party affiliations.

Kauffman [108] postulated that life itself is a critical state poised between the uncorrelated random state of complete chaos and the completely ordered state of death. The criticality effect induces the emergence of this precarious state called life at the 'edge of chaos' and this is where humanity struggles to survive.

4.2.9. Maximum Entropy Effect

The Principle of Maximum Entropy has been formally used as the fundamental basis of thermodynamics. The principle posits that the probability density associated with a given data set is the one that yields the maximum entropy, or equivalently, the minimum order for the underlying system. The empirical parameters determined by the experimental data constrain the possible choices for the probability density and determines the unique distribution that gives the minimum level of organization for the system consistent with the data. Suppose the data determines the average value of the physical observable. In this case the probability density is exponential and the average value or rate determines all the observable properties of the system.

Before the personal computer the landline telephone network satisfied the communication requirements of most people. The landline telephone traffic was estimated by the arrival of calls believed to be a set of mutually independent random events, with the duration of each call assumed to have an exponential distribution in time. The telephone traffic could therefore be characterized by the average rate of call arrivals, along with their average duration time. In one sense the world-wide telephone network was built on the assumption that this distribution and these two empirical parameters faithfully modeled the communications network. Thus, this simple model dominated telephone technology and the growth of communications for nearly a century. More complex models of communication traffic became increasingly necessary as the telephone web grew and we take this up in due course.

Another example of the application of the Principle of Maximum Entropy assumes the data determines both the mean and the variance of the observable. In this case the probability density resulting from entropy maximization is determined to be Normal and all properties of the system being modeled are determined by the empirical mean and variance. In this way the linear additive nature of the world of Gauss, Adrian and Laplace is consistent with the Principle of Maximum Entropy for systems in which the elements are essentially independent of one another, or at most only weakly dependent on one another. This is the intellectual world in which most people are comfortable. The events of such a world are not boringly predetermined while at the same time they are not overly capricious.

It is also possible to incorporate constraints beyond the algebraic ones usually discussed, that is, beyond the variable raised to an integer power [109]. This new constraint is established by the requirement that the probability density be scale free and the mathematical function that satisfies the scale-free requirement is the logarithm. One way to incorporated this observation regarding the logarithm is by using a technique introduced a number of times before and that is to transform the data. We logarithmically transform the original data and note that such a transformation emphasizes small sizes and de-emphasizes large sizes. Here again we place the simplest constraint and assume the average of the new physical variable is constant. Consequently the probability density that maximizes the entropy of the system is exponential.

The new exponential probability can now be transformed back to the original variable, that is, the exponential distribution is in terms of the logarithm of the original variable. Curiously enough the exponential and the logarithm are inverse operations of one another so that the distribution consistent with the underlying system having the maximum entropy (minimum order) and constrained by the logarithm (scale free) is that of Pareto. Put differently the exponential in the logarithm variable corresponds to the inverse power law in the original variable. This constraint on the logarithm ties widely separated scales together, resulting in the inverse power-law distribution.

4.3. RANK ORDER

I like to know the relative size of things. I am more comfortable when I see how things fit into the natural order of how I think they ought to be. My experience is that everyone else has the same predisposition, even if their cognitive maps are vastly different from my own. In post adolescence males the significant question to be answered might be: Who has had the most sexual liaisons? Even those that do not directly participate in such comparative discussions listen with a perverse curiosity. Towards the end of their college years this question might be replaced by: Who is the smartest guy in the room? As their careers begin to take shape it might morph into: Who makes the most money? Or in the case of an academic the concern might be who has published the most papers, or received the most citations or awards? Society often characterizes this wanting to compare, contrast and order events as a strictly male trait, reducing it to penis length comparisons. But woman have their own version of the same syndrome. Historically the female version might take the form: Who is the best dressed woman in the room? The ordering in the social hierarchy seems to have been very important to women at one time. Today there is probably less difference in the male and female points of view, except perhaps in what is being ranked.

In the contemporary world the tendency to rank order events is seen as a weakness of human nature, being something we should teach our children to avoid. But in a historical context, being able to run faster, distinguish between corresponding shades of color and sound more clearly, and plan more successfully for the hunt, all had survival advantage. Could it be that rank ordering individuals was part of the natural selection process? Those that ranked the highest were the most suitable mates and what we interpret now as a failing is actually a vestige of this once all important ability to insure the survival of the species. What is the evidence for this speculation?

When a data set is placed in ascending order from the smallest to the largest something remarkable often occurs. The size or rank ordering of the data follows an inverse power-law distribution. Notice that this is a different kind of inverse power law from that of income, say, because the size itself does not matter only its relative order, its rank. Consider a thick book written in English, such as

Steinbeck's *East of Eden* and have someone count the number of times each word was used in the book. It is not necessary to have someone do the counting, it has already been done. Not for East of Eden, but for a book of comparable length. In this count the number of times 'a' appears, and 'the' appears, and so on, are recorded. Rank one is given to the word used most frequently; rank two to the next most frequently used word, *etc*. The Harvard linguist Zipf noted that the product of the relative frequency of a word and its' rank produces a constant. This hyperbolic relation, the product of the two variables equaling a constant, implies that the rank-ordered distribution of English words is inverse power law, with a power-law index of one. This simple distribution of the rank order in the frequency of word usage is referred to as Zipf's law. Consequently, Zipf's law states that a rank-one word occurs twice as often as a rank-two, three times as often as a rank-three, and so on. A similar relation is found for other languages as well.

It is not just the frequency of words that has this kind of rank ordering distribution. At the turn of the twentieth century an empirical law regarding the rank order in the size of cities was discovered by Auerbach , determining that the product of a city population and rank are constant. Ordering the cities in a country from largest to smallest population, a city's rank is given by its ordinal position in the sequence. In this way he obtained a hyperbolic relation between city size and rank. Or in the jargon used here Auerbach's law is a rank-ordered distribution with an inverse power-law index that is slightly less than one.

It was just this kind of rank order relation that Willis found in his species size data relating to evolutionary biology. As Willis pointed out, the myotypic genra, with one species each, are always the most numerous; the ditypics, with two species each, are the next in rank, genera with higher numbers of species becoming successively fewer. The generalized hyperbolic relation consisting of the product of the relative frequency and the rank raised to a power is again given by a constant. The inverse power law was found to be true for all flowering plants, certain families of beetles as well as for other groupings of plants and animals. Given the empirical basis of Zipf's law, Auerbach's law and Willis' law, which are representative of a large number of other empirical laws, we ask the question: Why are rank-order distributions of the generalized hyperbolic form and not another?

Consider the difference between Pareto's law and Zipf's law. In the case of Pareto the relative number of people having a particular income level determines the distribution. To express the income data in terms of rank order we parcel the distribution into discrete income levels with the highest level being rank one, the second highest rank two and so on. This constitutes a change in representation from the relative number of people with a given level of income to the rank order in the level of income. The fact that both representations are inverse power law determines that the two power-law indices are simply related. In fact all power-law phenomena generate power-law rank ordering with characteristic power-law indices. Consequently there is a deep connection between the complexity of phenomena in the real world and their rank-order distribution.

So how does this relate to our inherent tendency to order things? I do not know, but it is certainly interesting.

I may not know how the rank ordering done by biological and social evolution relates to my comfort level in having things ordered in my cognitive map, but I do think it is related to how my brain organizes and orders things. My experience of events in the world are condensed in the mapping processes, and regardless of the representation used their relative ordering and perhaps even their relative separation in space and time is maintained. Thus, there is a resonance when the pattern in my mental map matches that of the world. That 'resonance' is captured in an author's use of language and the frequency with which certain familiar words are used, such that the story flows at an uninterrupted pace and it is a pleasure to lose oneself in the tale. Even the use of unfamiliar words if done infrequently are not disruptive.

The topics of complexity resonance, human response and inverse power laws are taken up in the next chapter.

4.4. WHY INVERSE POWER LAWS?

What a complex life lacks is balance, something that I was taught is school is desirable. Eat in moderation, so that only small amounts of exercise are necessary;

enjoy work but maintain an active family life; always balance one activity against another and avoid extremes. However such a view of balance overlooks the lesson of the simple lever. A child's see-saw surprises a child when they first see that their parent sitting close in to the focal point, can be balanced by them sitting far away from that point. If they lean far back their parent can even be raised high in the air. This toy is a metaphor for the real balancing observed in economic activities such as leveraged buyouts and transactions. The true balance in life comes from being able to exploit the intrinsic imbalance in the distribution of the things that are desired and successfully leverage it for an advantage. Of course, since people do not share the same value structure it is possible for two people to leverage one another for mutual advantage. If the distribution of talent and income were Normal then leveraging a piece of talent for a piece of wealth would not be possible. An asymmetry, or tail, is necessary for leverage. In a simpler time such exchange was the phenomenon of patronage, where a wealthy individual would support a talented person, while they gave expression to their art. This arrangement had some draw backs, but by and large it was mutually beneficial to the artist and the benefactor and eventually even to society. View Michelangelo's David the next time you are fortunate enough to visit Florence to assess the potential benefit to society of leveraging such talent.

The failure of simple parameters and Normal statistics to describe the behavior of complicated phenomena entails associating inverse power-law statistics with complexity. This association has guided the identification of mechanistic effects operating within complex networks, that result in the genesis of various inverse power laws. The entrepreneurial effect operating in an economic context leads to a Pareto tail in the distribution of income; the Matthews effect produces the inverse power law in the connectivity of the Internet; the allometry effect guides biological evolution in the generation of new species; and the contagion effect results in the intermittent spreading of laughter, rumor and disease.

It is interesting to note, as have others, that the mathematical theories developed to explain complex phenomena using inverse power-law distributions are continuously being rediscovered. We have given some indication of this in the discussion of the mechanisms producing power laws and this re-inventing has been documented in detail most recently by Simkin and Roychowkhury [110] who trace the historical roots of the ostensibly new concepts of preferential attachment and self-organized criticality and even demonstrate how the two dovetail.

The mechanisms presented in this chapter provide a preliminary understanding of how the complexity of a single complex network can come about. The complexity is the result of the loss of Normalcy. In turn there are multiple candidates for the cause of this loss: the loss of independence of the elements; the dynamics of the individual elements becoming significant (perhaps chaotic); the emergence of a mutual reciprocal influence of the complex elements, due to changes in the range of the force law; or any combination of these. In short, complexity is how we experience phenomena when for one reason or another the central limit theorem breaks down and enhanced variability takes over. The one or two parameters characteristic of simple networks are replaced by a distribution of possible parameter values. It is this loss of the simple mental model and our reluctance to replace it with a less desirable, but more realistic power-law model that engenders despair about complexity.

CHAPTER 5

Complexity Management Principle

Abstract: Science is about finding order in the panorama of the world and embracing a perspective that includes the falling of apples and the motion of planets; the behavior of the individual and the actions of groups, large and small; the information content of an encyclopedia and the Wikipedia; in short, science does not, and should not, have any boundaries with regard to content. The terrestrial and the cosmic are part of the give and take in science, with the goal of uncovering the principles and laws that determine how the universe functions, along with the individuals within it. For most people, science appears to be separate and apart from the world in which they live. The principles and laws of science do not seem to apply to the general interactions among people; due, in part, to the fact that principles have not been found for everyday decision making; laws have been notoriously absent from mundane thinking; rules have been sought in vain in the growth of society; and indeed canons go begging in the multiple complex phenomena within the human sciences, despite over two hundred years of effort to either invent or find them. In this chapter we examine the Principle of Complexity Management, whereby a system with greater information, but perhaps lesser energy, can dominate a system with lesser information, but greater energy. The principle is a recently proven generalization of an observation made by the mathematician Norbert Wiener, and may be one of these universal principles.

Keywords: Complexity management principle, Global warming, Habituation, Inverse power law, Laws, Leaky faucet, Memory, Network-centric warfare, Universality Norbert, Wiener.

The goals of economists studying global financial markets; the objectives of sociologists scrutinizing terrorist organizations; the intent of neuroscientists to understand neuronal networks and the aims of others in the human sciences, are no less worthy than those pursued by the physical scientists. Some would argue that understanding these non-physical phenomena are, in fact, more important than are their physical counter parts, because they are so closely related to the every day events of people. Therefore, it is exciting that even though the muse has not pulled back the curtain on the principles of human science, it appears that the first glimpse of scientific laws of human conduct, may have been made.

The recent surge of interest in networks and the effort to develop a network science has provided a foundation for general principles outside the physical sciences. It should be understood that network science does not yet exist. There are various theories and models, but nothing that has risen to the level of science that is distinctly dependent only on networks. Network science is a hope for how science may develop in the future. It should be clear that if network science is to exist, it will stand outside the traditional disciplines, in recognition of the common aspects of various disciplines coincident across complex phenomena. The common features across phenomena result in the interchangeability of their network representations. Inverse power-law statistics and pink (1/f) noise capture complementary aspects of complexity in dynamic behavior through the mechanistic effects discussed in the last chapter. Without identifying the common features, it would seem unreasonable to compare music to a dripping faucet, or an oil spill to the milk in a cup of hot coffee.

The first step in formulating a principle of network science has been taken, through the collective scientific insights of what constitutes a complex network, provided by scientists over the past decade. The next step in formulating a principle for understanding the complexity of human phenomena is to explain how complex networks influence one another. The influence can be deterministic by means of direct physical control, as in leaning a motorcycle through a turn, or trying to convince your teenage son or daughter that having sex, whether protected or not, has dramatic unintended consequences. On the other hand, the influence can be statistical, where the fluctuations in one network gently perturb those of another, producing an unexpected variety of responses. The transfer of information by weak coupling of one network on another, may be absorbed by the second network resulting in no long term response. On the other hand, the influence may be cumulative and the perturbing network may eventually completely dominate the network being perturbed. The transfer of statistical influence through information is more subtle than that through energy perturbation, but may be more common in the interacting complex networks that nature designs, than are the deterministic controls designed by engineers.

A traditional way to frame our understanding of the question of how complex networks influence one another is to construct the separate equations of motion for the two networks and then allow them to couple, usually through a linear term. An engineer would design a coupled system in this way and therefore this is the first way science tries to understand how nature does it. If this does not work, we can always go back to the empirical approach; that being, to put the two networks of interest into direct contact with one another and record what happens. Of course, the empirical approach is almost always costly, often not feasible, and, when dealing with human beings, is frequently limited by considerations of ethics. However, nature is filled with examples of species that feed off one another, when they are allowed to freely interact, as we discussed earlier.

Before turning to the social context let us look at how individuals react to various kinds of stimuli. The initial shock to the skin of the icy water of a stream is attenuated to a warm tingling sensation; the strong odors that command attention, soon fade, leaving an undetectable scent; the annoying traffic noise outside the motel window is replaced with the comforting quiet of sleep. All these stimuli start clearly in consciousness, but in a relatively short time their influence on what is experienced dissipates. This is the phenomenon of habituation in which a simple stimulus first attracts attention and then relinquishes it over a relatively short time, without the stimulus changing form. Therefore, people fall asleep in front of the television, or reading a book, or in the middle of reading a scientific paper. Habituation is examined here in terms of a complex signal exciting the brain, using a suitable function to measure the level of response of one complex network being stimulated by another. This examination reveals how complexity, in the form of information, is transferred and controlled by complex networks. My colleagues and I call this transfer and control of information the *Principle of Complexity Management* [71]; it is interesting because the complexity of the sender and receiver are crucial in determining how much information is actually transferred.

One thing that has become apparent in discussing the transfer and control of information, between complex networks is the importance of pink noise and inverse power-law variability. The old question regarding whether nature or nurture is more important in determining the character of individuals is addressed in the context of interacting complex networks, with one network being the human brain. A related question concerns to what extent aesthetic choices are a balance, between biology and sociology. Some recent studies may be able to provide tentative answers to these and other similar questions.

5.1. UNIVERSAL LAWS AND PRINCIPLES

Stories in the morning newspaper, when done properly, answer the questions: Who?, What?, Where?, When? and Why? A good story does this with the artful presentation of data and the skillful revealing of patterns hidden within the data; the patterns contain information that is used to develop the why. The why is often used to imply causality, but as Aristotle explained, there are different kinds of causality. What is absent from journalism is the why associated with scientific causality. Why the information has the pattern it does is not discussed, unless that pattern is part of an even larger pattern. The why of things is the province of science, where the scientific method enables scientists to extract knowledge from information, that is, to draw conclusions. Scientific theory has historically been excluded from news stories, since all too often it appears as opinion masquerading as knowledge. Consequently, scientific theory is replaced by ideology and speculation replaces inductive conclusions, which belong on the opinion/editorial page of the newspaper, or in a blog, not as part of the news.

A large part of human thinking is not done through the application of knowledge, but through comparing patterns; a process of comparing information stored in the brain with information being supplied by the environment. Humans continuously make decisions based on the compatibility of these patterns, rather than on their knowledge as to why these patterns arise. A batter does not hit a ball because of his knowledge of Newton's laws; he hits it because the pattern of the incoming ball matches his mental picture of a home run. In the same way a fielder does not calculate the parabolic trajectory of the fly ball, but matches the observed path of flight, with his mental picture, and races with an outstretched glove to where the two trajectories coincide in space and time.

If we make everyday decisions based solely on comparing internal to external patterns, then what is the value of knowing why and having a scientific theory? Perhaps the merit of science lies in the fact that the patterns generated in a variety of ostensibly different situations may be synthesized into universal forms. The stories of the founding of modern science, with which we started this book, had to do with Netwon's formulating such universal laws and principles from experiment. His first principles had to do with the laws of mechanics and the definition of mass that was supplied to him by Galileo. Once these rules of physics were verified and implemented, their implications were determined by mathematical reasoning. Therefore, postulating a universal force law for gravity allowed Newton to put celestial motion into the framework of mechanics and predict the orbit of the Earth around the Sun, the Moon around the Earth, and all the rest of celestial mechanics. This success became the scientific model that was so strongly emulated in the nineteenth century.

The importance of laws and principles in science cannot be overstated. Thus, in order to appreciate the significance of a new one, with a candidate such as the Principle of Complexity Management, let us discuss a central concept in physics and biology and a basic law associated with it as an exemplar. The concept is energy and the law is the conservation of energy. The total amount of energy does not change, regardless of what nature does. Energy is a sophisticated concept and not easily explained, so we follow the lead of the Noble Laureate physicist Richard Feynman and present an allegory to aid in understanding its conservation.

In my allegory I draw on the fact that I had six brothers, two older and four younger, and in a large family everyone is assigned certain daily chores. In the parable I am given the task of keeping track of a set of 30 absolutely indestructible blocks my youngest brother received on his birthday. The featureless cubes are completely equivalent to one another. My brother takes the set of blocks into our bedroom at the beginning of each day to play. In the evening I go into the bedroom to put the blocks away in the toy box. Over a period of time I notice that there are always 30 blocks. To check this observation, each day I count the blocks and each day I find 30. Then one day a block is missing, the count is only 29. With some looking around I find the missing block under the blanket on the bed. After that I scan the bedroom to make sure that the number of blocks remains fixed.

A few days later the count is again short - with only 28 blocks. The two blocks are not in the bedroom, but I notice the window is open and looking outside I see the two errant blocks on the lawn. On yet another day, my count was 32. Needless to say I was concerned, until I remembered that my brother had a friend from next door over to play, who had brought with him an identical set of blocks. The neighbor had left two blocks when he went home, so I sent them back next door, made sure the window was closed, and did not let my brother's friend back in the bedroom (remember this is only a story). Everything was going fine until on another day I count the blocks to find 25. However, I notice the toy box and go to open it, but my brother invokes his right to privacy. He would have told our parents, so I could not open the toy box.

I was curious and being of a scientific inclination even then, I invented a procedure for determining if there were any blocks in the toy box, without opening it. I weighed each block and found they weigh exactly 4 ounces each. I then weigh the toy box at a time when I can see all 30 blocks. The next time I want to determine if the toy box contains any blocks I weigh the toy box again, subtract the weight without any blocks, and divide by 4. The number resulting from this calculation subtracted from 30 is always exactly equal to the number of blocks I can see. This is an 'empirical law' regarding my brother's blocks. The number of blocks I can see, plus those determined from the weight of the Toy box, is always equal to 30.

This equation works fine for a while, but then I notice another deviation. One evening, after my brother's bath, I observed that the dirty water in the bathtub had changed its level. My brother had been taking blocks into the bath and leaving them there. I could not see the blocks through the dirty water. I decided I could determine how many blocks were in the water by adding another term to the block conservation formula. The original depth of the bath water was 12 inches and each block raises the water level by one-hundredth of an inch. The new formula for the total number of blocks contains a term given by the present height of the bath water, minus its initial height, divided by one-hundredth of an inch. The new empirical conservation law is:

observed + # in toy box + # in bathtub = 30.

The first term in this law denotes the direct observation of blocks and the rest of the terms infer location of the blocks by examination of the toy box weight and bath water height. In the gradual increase in complication of the world of blocks there is a sequence of terms, each representing a different way of calculating how many blocks are in places where we are not allowed to directly look. As a result there is a complicated formula for a quantity that has to be computed. The result of this calculation must always yields the same number, otherwise it is not a conservation law.

This block counting allegory reduces to the conservation of energy when, following Feynman, I make the further stipulation that there are no visible blocks. Consequently, we disregard the first term in the empirical law, so there is no direct observation of the blocks; only the inferential terms remain. This transforms the blocks to an abstraction that can only be determined indirectly by computation. This is the situation with energy, it is an abstract quantity that can be neither created nor destroyed, and mass is another form of energy as Einstein's law $E = mc^2$ reveals; m is mass, c is the speed of light and E is the equivalent amount of kinetic energy. But energy is more than an abstraction. I feel the sun on my face when I step outside. I see the shifting patterns of light and dark in my yard from direct sunlight and shadows. I hear the birds gathering around the feeder squabbling over the seeds and smell the flowers in the bed below the feeder. All the senses reveal energy in its multiple interconnected forms. This universal concept enables understanding of the various phenomena in terms of physical processes. One principle can do all that.

The conservation of energy is contained in the first law of thermodynamics and, as in the allegory, cannot be proved, only disproved. This law of the conservation of energy is a summary of vast amounts of experimental data and its implications provide deep understanding into the dynamics of physical systems. The energy in a simple mechanical system takes two forms, kinetic and potential, and the total energy is constant in the absence of friction; friction being a mechanism that transforms mechanical energy into heat. Consequently, the arm of a pendulum has maximum speed at its lowest position and minimum speed at the end of its swing.

The energy is continuously shuttled from all kinetic at one extreme to all potential at the other extreme and back again. The potential energy contained in the ice, snow and rocks at the top of a hill is converted into kinetic energy in an avalanche. In the valley below, there is no place to deposit the kinetic energy of the cascading mass so it is destructively absorbed by the trees, landscape and buildings. The abstract energy, that once resided in the scenery of the hillside, becomes the energy required to bend, break and destroy the quite valley. Knowing this one law, therefore enable the scientist and engineer to calculate the dynamics of every kind of mechanical device. The motion of any particular mechanical device can therefore be understood using this one general principle.

It is, therefore, not unexpected that scientists would want to find the same kind of abstract principle for the human sciences. Such a principle may, in fact, have been found involving the second law of thermodynamics, not the first. In one form, the second law of thermodynamics states that isolated physical systems become more and more disordered with the passage of time. The decrease in order is measured by an increase in entropy. The entropy increase is the result of the dynamics of a thermodynamically closed system, which gradually loses information over time. The loss of information is due to the fading influence of the initial state of the network on the evolution of the network, until all influence of the initial state is gone. Once I am in Boston traffic it does not matter whether I started my adventure in Cambridge or Concord; the initial state is forgotten, and I must attend to the present situation in order to survive. The increase in a network's entropy measures the decrease in the network's information and the erasure of memory.

The phenomenology of thermodynamics ought to be derivable from mechanical principles and attempts to do this are the subject of statistical mechanics. In statistical mechanics the idealized laws of mechanics are made more 'human' through the introduction of uncertainty, which is to say, through statistical fluctuations. In this way thermodynamic information is transformed into statistical mechanical knowledge in which we know that all patterns have noise. The uncertainty of noise is understood to come from the environment in simple networks, but can also be generated by dynamics in nonlinear networks. Consequently the information in simple networks is found to be constant or to decay due to the unknown influence of the environment. The patterns in complex networks can vary over time even being periodic without being harmonic; more like a roller coaster than a pendulum.

Very often what people mistakenly call noise is not just pattern-free random variation. In complex networks, such erratic appearing variation may contain the very information investigators are seeking to understand and predict. The signal variability in medicine might be the subtle change in gait, that is diagnostic of neurodegenerative disease, or the understandable mood swings of the soldier returning home from the middle east, might be indicative of post traumatic stress disorder (PTSD), or the shifts in color or sound could be the aesthetic appeal of a piece of art. What science seeks to understand is how to quantify and predict the change in response of a complex network to small changes in stimuli. A simple measure of this sensitivity is the cross-correlation function, which quantifies the level of network response to stimuli.

5.1.1. Measures of Influence

Physical scientists are confident in their predictions, because a great deal of their research is associated with establishing causation and its consequences when studying a phenomenon. It is not sufficient that one event precede another in time, causality requires a material relation between the cause and the effect. The batter causes (produces) the home run by imparting part of the momentum of his body to the bat, and from the bat to the ball with a thunderous "wack". But most phenomena lack this clarity, or singleness of purpose. The very notion of causation fades from sight in considering complex processes, such as the beating of the human heart. However rather than pursuing this discussion of cause and effect further, it is sufficient to recognize that in the human sciences the role of causality is traditionally replaced with that of correlation. Causality ought to play a central role in any basic theory of human science; but in the present state of phenomenology only the correlations between human events are typically understood and not their casual connections.

Fig. (**5.1**) displays the inter-beat time interval in 100 consecutive heart beats, taken from a much longer sequence. This is a series of the time intervals between events, in a sequence of events. It is evident from the figure that although the heart rate is not constant, the time interval does not change radically; it has an average time interval of 0.8 sec between beats. The variation around the average heart rate is approximately 7%. The standard deviation in the time series is 0.054 sec, which seems to support the notion of normal sinus rhythm, since 0.054 << 0.80. The question presents itself: How much information could be contained in these random changes in the time intervals between beats?

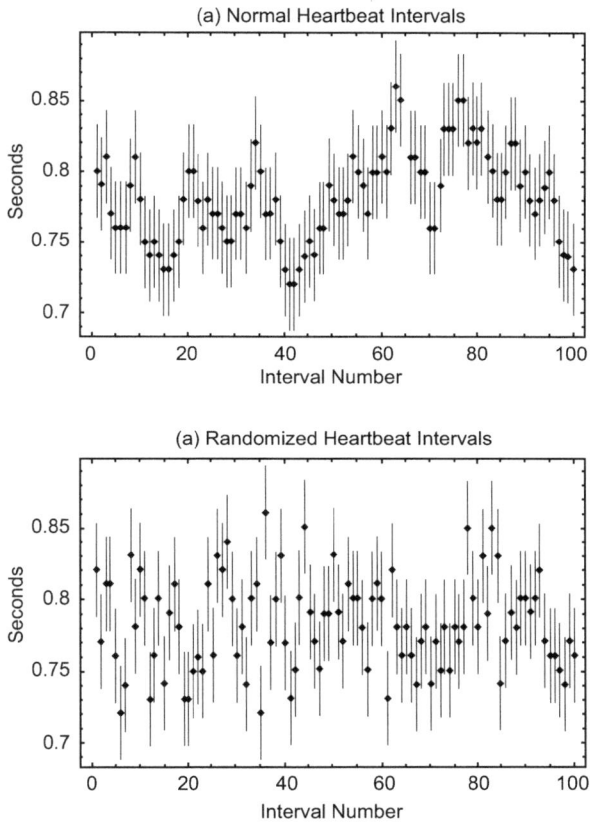

Fig. (5.1). (a) The time series for the interbeat intervals of a normal healthy individual lying supine is given by the dots and the bars indicate plus and minus the standard deviation. (b) The same data as in (a), but randomized with respect to the beat number. Therefore characterizing the beating heart by a single number, such as an average heart rate appears to be inadequate.

The time series in Fig. (**5.1a**) denoted by dots appears to be a sequence of random numbers, hopping between values in a capricious way. Is there an underlying structure that correlates what happens at one beat to that of another? A qualitative answer to that question is given by a simple transformation of the data; randomly shuffle the time intervals. Shuffling destroys any underlying relation between beats that is a consequence of their order in the sequence, that is, any correlations in the original data sequence are lost in the shuffled data. This test for randomness is depicted in Fig. (**5.1b**). The shuffled data has the same average and standard deviation as the original data set. This conservation of average behavior occurs because no time intervals have been added or subtracted in the shuffling procedure, the data points have only been randomly moved around relative to one another.

Comparing Fig. (**5.1a**) with (**5.1b**) it is evident that certain patterns in the upper graph such as the monotonic increase or decrease in heart beat intervals over short times do not appear in the lower graph. No regularity in the time interval data survives the shuffling process. The data shown is typical of normal healthy young adults. Observing the variability in the data, it is evident that characterizing the cardiac cycle with a single number, such as the heart rate, ignores any underlying patterns of change within the heart beat data. As we mentioned, patterns contain information about the measured process, and by ignoring such patterns in physiologic time series, information is discarded that may have value. It is quite possible that information contained in the variability pattern may be more important than the average values so carefully taken by every admitting nurse in every hospital in the western world.

There appears to be a qualitative difference between the original beat-to-beat data and its shuffled version. So how do we use our modeling skills to quantify this difference. One technique that has gained support over the years is the correlation function, which is a generalization of the variance of a time series. Consider a second example consisting of a sequence of the time intervals from one heel strike to the next heel strike of the same foot; the stride interval. A stride interval time series consisting of 500 steps is depicted in Fig. (**5.2a**). From these data one might guess that the random fluctuations are riding on top of a periodic signal. The autocorrelation function measures how one part of a time series influences subsequent parts of the same time series. This is done by first adjusting the data by subtracting the mean value from each of the data points and dividing by the standard deviation of the time series. Consequently by taking the product of the adjusted time series with itself and dividing by the total number of data points yields unity. This is the first point in Fig. (**5.2b**) and indicates perfect correlation of a process with itself. Shifting the second data set in the product by one unit of time, called the interval lag, the value obtained determines the average influence of all data points on each subsequent data point. It is evident from the graph on stride correlation in Fig. (**5.2b**), that as the data points in the two series separate, their influence is diminished for the first five seconds and approaches zero. However, after further lagging the two time series the autocorrelation begins to increase and has a value of approximately 0.25 at a lag of ten seconds, after which, it begins to decrease again. This decrease in correlation followed by an increase means that elements of the data that are ten seconds apart influence one another more strongly, on average, than those that are closer together in time, but the effect is not monotonic it is periodic. The stride interval data has a ten second period, weak modulation, with a diminishing amplitude in time on average.

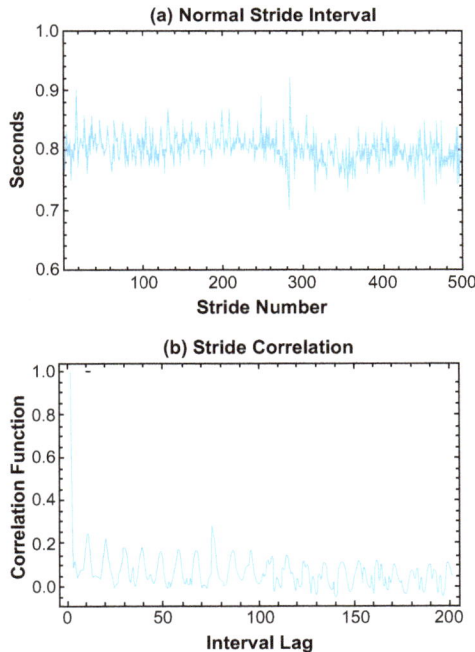

Fig. (5.2). Part of a normal stride interval time series is given in (a). In (b) the autocorrelation fundtion of this time series is depicted.

Consequently it is not a simple matter to predict the kind of correlation a time series will have. In fact looking at the data from a time series, without a quantitative measure, it is essentially impossible to perceive what intrinsic structure might be present. Of course if we shuffle the data this interdependency of the stride intervals is lost, just as it was for the correlation in the heartbeat data. How this correlation changes with various pathologies is the concern of researchers, who are studying the influence of the body's nervous system, that tells the heart when to beat and the leg when to strike out. The subtle variations in the nerve impulses are recorded in the variations in physiologic time series, so that the correlation functions can guide the diagnosis of the medical practitioner.

Up until now, we have been concerned with the behavior of a single complex network and what can be learned from the influence of that time series on itself over time, an auto-regulation effect. The auto-regulation of the cardiovascular and motor control systems can be evaluated from their individual time series. If what happens at late times, becomes completely independent of what happens at early times, then the network can only self-regulate for short times. Self-control is a matter of knowing how far the present data is from historical data and adjusting the network to bring the two into alignment. When auto-regulation breaks down, the result is pathology or disease; the modern interpretation of disease is the loss of variability in the physiologic time series, and is interpreted to reflect the loss of complexity within the physiologic network [46].

However, if the question is how the cardiovascular system influences the motor control system or vice versa, that is something all together different. A new measure is used to determine the influence of one network on another. The question is how to determine if there exists a statistical relation between the two: Does one data set depend on the statistics of the other or not? It is always possible to construct such a function from data and measures how the variations at a point in time in one network, P, influences the behavior at another point in time of a second network, S. Consider the time series generated by the network S to consist of a set of unit amplitude pulses of random sign, which we call a telegrapher's signal. Like the signal of the now obsolete telegraph message, the key is either open or closed for various lengths of time. It is the distribution of these intervals that contains the information of interest. In simple cases the information is encoded in messages. In more complicated cases the embedded information is masked by random variations in the time intervals. In this way the values of the time series randomly fluctuate between +1 and -1. A typical data set would then look like the time series shown in Fig. (**4.6**). The information is contained in the time intervals between switches from plus one to minus one and back again. The switch from one state to the other is called an event and we switch focus from continuous dynamics to discrete events.

It is clear that if the events are truly random, that is, completely independent from one switch to the next, then the average value of the time series is zero for a sufficiently long sequence, since a non-zero average value would constitute a bias or correlation. Note that the average is determined by adding up the data points and dividing that sum by the total number of data points. Such series have been used to represent the firing of neurons, the arrival of messages in a communications network, the typical height of individuals in a Normal population, and a host of other phenomena focused on the counting of discrete events.

If we have a similarly defined data set for network P we can take the product of data points at each tick of the clock, one data point from S and the other from P, to obtain a product data set of the same length as the individual data sets. The average of this product function has three important limits; zero when the two data sets are independent of one another; one when the two data sets are essentially identical; and minus one, when one time series is always in opposition to the other. The name of the normalized product is the cross-correlation coefficient and it can assume all values between minus one and plus one indicating the degree of dependence of one time series on the other. The attempt to construct functions or processes that determine the influence of one complex network on another is probably as old as civilization itself. In medieval Europe it was not so straightforward to distinguish science from pseudoscience, say astronomy from astrology.

This overlapping of astronomy and astrology may appear silly from today's perspective, but many respected scientists in the seventeenth and eighteenth centuries were astrologers. After all, they did have to eat. Johannes Kepler, whose work was one of the pillars supporting Newton's celestial mechanics, was mathematician, astronomer and astrologer. He was court mathematician (1600-1612) to Emperor Rudolph II and one of his duties was to provide astrological advice and cast horoscopes for the emperor. He had been casting horoscopes since his student days at Tübingen. The intertwining of astronomy, astrology and theology was common at the time and provided structural support for the cognitive maps of all three. Moreover, the fact that Newton, the father of modern science, was an alchemist, should neither be forgotten, nor should it detract from his scientific contributions. Of course, we no longer have emperors, nor do scientists tout astrology or alchemy, but the model for giving scientific advice at the highest levels of government for suitable compensation, has not changed significantly.

The cross-correlation is defined in terms of data by displacing one time series from the other, for a specified time lag, to determine the time span of influence from each point in the data set. However, it is not easy to determine the theoretical form of the cross-correlation function from the dynamics of the interacting networks. In general we cannot know the formal structure of the cross-correlation function, unless we know the full dynamics of both networks. For example, if the network P has random fluctuations with Normal statistics and network S has a dissipative dynamic response Fig. (**5.3**) depicts the decaying average response of network S to the random perturbations by network P. The response becomes exponentially weaker with the time separation from the point of application of the random stimulus in time. A number of other simple situations can be described by tractable models, but this is not generally true. Typically the explicit form of the cross-correlation function is not available.

Recently, however, a general form of the cross-correlation function was constructed for the response of S to a weak stimulation by P. The mathematical arguments rely on the composite network satisfying only a few conditions. What is most interesting is that the value of the cross-correlation function quantifies how responsive the complex network S is to perturbation by the complex network P.

The first condition is that both networks are complex in the sense that their random fluctuations are not described by Normal distributions. Both networks have fluctuations described by inverse power laws and the degree of complexity is quantified by the power-law indices. Since we began this section with time series, the inverse power-law distribution represents the time interval between events, whether it is the time interval between the beats of the heart or between steps in a walk, or the firing of neurons in a neural network. The inverse power law in time implies that the fluctuations in the time intervals between events are intermittent; they occur in self-similar bursts.

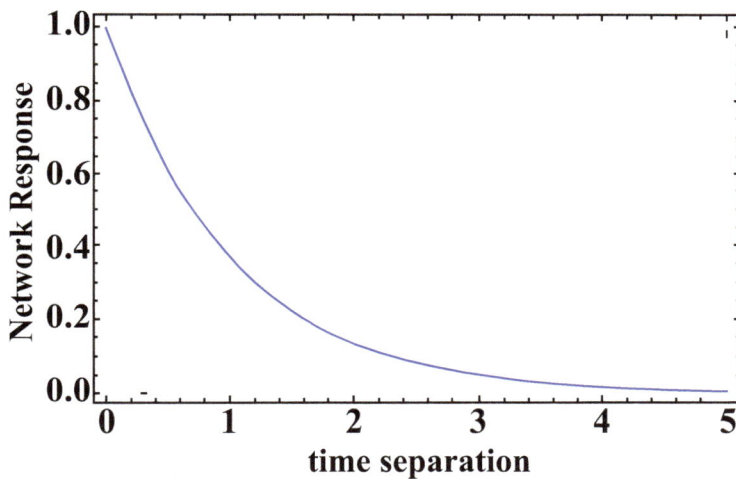

Fig. (5.3). The response of a dissipative network S to a fluctuation of perturbing network P is indicated by the decaying exponential. The time separation is in units of the relaxation time and the cross-correlation function starts at the value of one.

The second condition is that the stimulus is sufficiently weak that the response of S to the stimulation by P is linear. It was previously pointed out that linearity is not always a bad thing, particularly in the right representation. The linear response of S to P is not a single term, since both networks are dynamic and the response is not instantaneous. It takes time for the perturbation to wend its way through the multiple random interacting elements within S. The linear response function is actually a sum over the influences from perturbations occurring from farther and farther in the distant past and are of diminishing strength. The total strength of the influence on what is being measured in present time is therefore dependent on how fluctuations in S are internally related in the absence of the perturbation. This internal connectivity within the network S determines how that correlation carries perturbations forward in time. The first condition requires linear response to be the proper description of the coupled dynamics of the two networks in which the linear response function is the probability of a perturbation at one time influencing an event within S at a later time. It is worth emphasizing that the linear response condition does not imply linear dynamics on the part of either network. In fact if the dynamics of either network were linear, then complexity in the sense used here would not exist.

5.1.2. New Principle

The complex network of interest S is assumed to have fluctuations, described by an inverse power-law distribution, with a power-law index μ_s, whose value lies in the interval between one and three. When the index is in the interval one to two the averages over time and averages over the probability density, need not be the same, in which case, the network is non-ergodic. On the other hand, when the index is in the interval two to three, the two ways of averaging yield the same result and the network statistics are ergodic. The assumption that a complex system is ergodic is made throughout the physical sciences, until quite recently.

The perturbing complex network P is assumed to have a power-law index μ_p ; in the same total interval as that in S, and the same interpretation in terms of ergodic and non-ergodic statistics apply. Consequently, the cross-correlation function obtained from the average overlap of the two time series depends on the two power-law indices as depicted in Fig. (**5.4**). The figure is constructed from analytic expressions for the cross-correlation function for specific values of the power-law indices as time series become infinitely long [70]. These expressions are obtained using a generalization of linear response theory, as discussed in detail elsewhere [70]. The figure depicts the cross-correlation cube asymptotically in time, which is a collection of all the values of the cross-correlation coefficients for different values of the indices for the stimulus and response. The height of the cube is the strength of the correlation, between the dynamic variables of the two networks and is bounded from below by zero and from above by one.

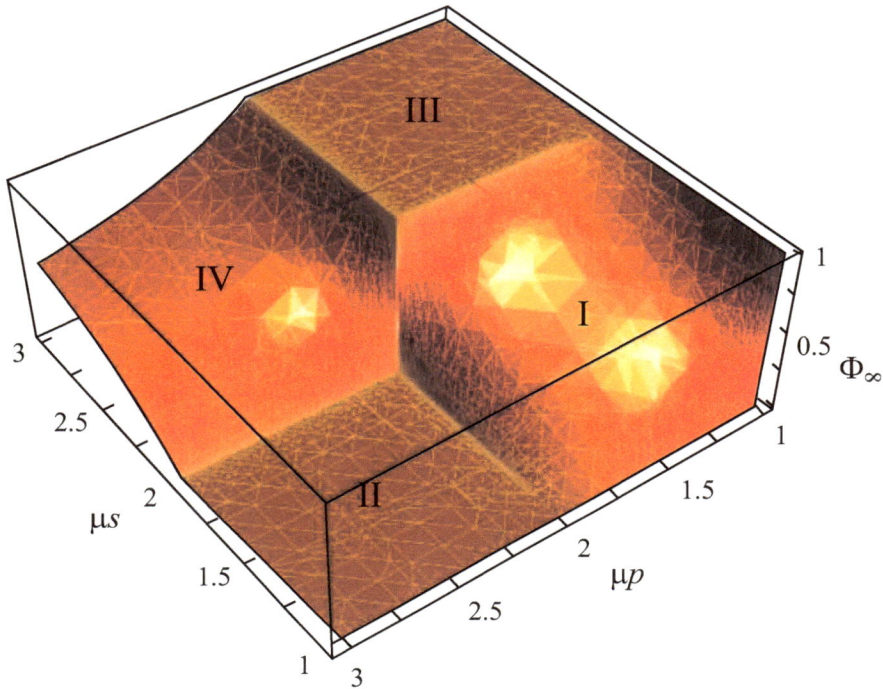

Fig. (5.4). The cross-correlation cube depicted here quantifies the asymptotic influence of one complex network P, on another complex network S. The responding network S has power-law index μ_s and perturbing network P has power-law index μ_p. The magnitude of the asymptotic cross-correlation coefficient clearly depends on the relative values of the power-law indices.

The plateau regions of the cross-correlation cube, labeled II and III, are particularly interesting. The upper plateau, region III, corresponds most closely to the Wiener Rule introduced in Chapter Four, in which the effect of P on S is dominant. The lower plateau, region II, apparently contradicts the Wiener rule, since there is no residual influence of P on S. But, as we shall see, this is not the case, and it is a direct result of linear response theory on which the cross-correlation cube is based.

How does linear response theory work? To answer this question, let us consider some simple physical examples. Water resists my efforts to swim rapidly, wires heat up when electricity flows through them and apple pies can take a prolonged time to cool after coming out of the oven. Each of these physical phenomena can be quantified using linear response theory and because they are simple, the descriptions can be reduced to distinct parameters. In swimming it is the viscosity of the water that ultimately determines how hard I work in swimming. Electrons in a wire encounter resistance due to the wire's atomic structure and the kinetic energy of the flowing electrons is converted into heat. The efficiency of transporting heat to the surface of the pie, where it is cooled by the air, is determined by how quickly heat is transferred from apple to apple within the pie's interior. Different materials transport mass, charged particles and heat differently and the average effect is determined by an empirical constant called a transport coefficient; viscosity, resistance and thermal conductivity in the examples given. Physics is replete with such phenomenological constants that dominate people's lives, from the coolness of the clothes they wear, to the efficiency of heating their homes, to how many miles a car can travel on a gallon of gasoline. But these are simple things that can be understood, more or less.

In the more general situation the network S in isolation has a large number of interacting elements. These elements could be people at a cocktail party, individual computers tied together on the World Wide Web, the interconnected neurons in the brain, geese flying in formation, Buffalo thundering over the mid-west planes, or any of a large variety of networks. The networks are sufficiently complex that if the network variable of interest were represented by a light bulb, it would flicker in an unpredictable way in response to the stimulus. The light would be on for an indeterminate length of time and then switch off for no apparent reason, only to remain off for another intermittent interval of time. In short the fluctuations in brightness would have a pink spectrum. An election district might have this form, with the variable of interest being the total vote for or against a controversial candidate, at a given point in time. A blinking quantum dot is a physical example of such a flickering phenomena and like the complex networks introduced above the probability density function, determining the switching time for the quantum dot, is inverse power law. I mention the quantum dots to indicate the span of scales in space and time over which these arguments have been found to be valid [88].

What response does the cross-correlation cube predict, such dynamically rich networks will have to weak stimuli? In physics we know that for a sufficiently weak perturbation a linear response description of the perturbed network captures the dynamics. This theory identifies phenomenological parameters such as thermal conductivity, that measure how rapidly heat is transported within a material; electrical resistivity, that determines the opposition to the flow of electricity in a material; with similar operational definitions for the coefficient of thermal expansion and so on. However such empirical parameters have not been developed within the human sciences and could be quite useful if they are found to exist.

The cross-correlation cube in Fig. (**5.4**) indicates that S can respond to P in a variety of ways depending on the relative values of the power-law indices in the perturbing and responding networks. Plateau region II indicates that the two networks are independent of one another after a long time. By contrast plateau region III indicates that the two networks are tightly correlated with one another after a sufficiently long time.

Consider plateau II where the range of values of the perturbing power law index μ_p indicates that the fluctuations in the perturbing network P are ergodic and have relatively short memory. The short memory is a consequence of the fact that the power-law index, between two and three, implies that the time interval between random events is restricted to relatively small values. In this case the average time between events is finite and the average time is a crude measure of the perturbing process. This is to be compared with the range of values of the response power-law index μ_s on this plateau, that is between one and two, and indicates that the fluctuations in the responding network S are non-ergodic and have a relatively long memory. For the response network the average time between events diverges and there is no characteristic time scale for the response process.

Thinking about this latter interaction it becomes clearer that the fluctuations in the responding network do not have enough time to adjust to the weak short- lived perturbation, before another perturbing fluctuation is generated. Because S does not have sufficient time to adjust to the old perturbation, before the P dynamics produces a new one, S behaves independently of the perturbation asymptotically in time. Consequently the perturbation has no lasting influence on the dynamics of network S and the asymptotic cross-correlation value is zero.

The opposite effect is observed on plateau III where the range of values of the perturbing power-law index μ_p is between one and two and indicates that the fluctuations in the perturbation have relatively long memory and no characteristic time scale. The range of values of the response power-law index μ_s on plateau III is between two and three and indicates that the fluctuations in the responding network S have a relatively short memory since the average time between events is finite. The fluctuations in the responding network have sufficient time to adjust to the weak long-lived perturbations and asymptotically the response loses its own identity and adopts to the statistics of P. Consequently the perturbing network P asymptotically dominates the dynamics of network S and the cross-correlation value is maximum on III. This is where the Wiener Rule applies and the weak perturbations of P dominate the response behavior of S.

Enough of abstraction and generality. What can we learn about everyday phenomena using the cross-correlation cube and the Principle of Complexity Management?

5.2. APPLICATIONS OF THE NEW PRINCIPLE

Human beings adapt. The loud din of a cocktail party, that strikes me on entering the room, fades into the background after a minute or so, and then I can enjoy a conversation at a reasonable volume. In a short time I am able to direct my attention and distinguish between individual voices and background noise. The hot bath water seems to cool even as I lower myself into it, but it is not the temperature of the water that changes so quickly, it is my body's sensing of it. The sharp pungent odor of sweat permeates training camps, but soon after stepping off the bus, the olfactory systems adjusts and the smell vanishes from consciousness. These are all exemplars of habituation; a phenomenon that is intimately related to how complex networks absorb and respond to information, when one of the complex networks is the human brain. This is our first application of the Principle of Complexity Management through an interpretation of region II of the cross-correlation cube.

A second example again concerns the functioning of the human brain, not during habituation, but how it functions when it is attentive to an external stimulus such as the sound of splashing water from a leaky faucet or the orchestrated consonance of classical music. In this application the brain does not habituate to the stimulus, but instead fixates on it. This is captured in region III of the cross-correlation cube.

A third example of the application of the cross-correlation cube, that probably could not be farther apart in scale and substance from the first two examples, is how the Earth responds to fluctuations in its heating by the Sun and the possible implications for climate change. Here again the response of the Earth to perturbative heating by the Sun appears to be in region III of the cross-correlation cube.

5.2.1. Habituation

People are continuously inundated by new sights, sounds, sensations and smells.If each sensory input was given the same attention a person would be over-whelmed, curl up in a ball, and tune out the entire experience. However, the brain does not give each input the same attention. Over the millennia humans have evolved a kind of sensor selectivity, one that automatically chooses which sensor signals contain useful information and which are merely noise. At one time the selection criteria probably contained survival advantage and although the criteria is now hard wired into our nervous system, through biological evolution, it may be atavistic. One sensory selection mechanism humans and other animals have evolved is that of habituation.

Habituation is a form of learning used by animals, including humans. It is ubiquitous and extremely simple way to learn to ignore stimuli that are no longer novel and consequently have no survival value. This mechanism prevents sensory overload and allows individuals to attend to new stimuli in a noisy background. Habituation can occur at different levels of the nervous system. Sensory networks actually stop sending signals to the brain for repeated stimuli. This effect occurs, for example, with strong odors. The west side of Buffalo was predominantly Italian when I was a boy. The corner stores had barrels filled with Dill pickles, huge cheeses suspended by webbing from the ceiling, sawdust on the floor and crates filled with live chickens. The smell would draw people in from the sidewalk and sixty-five years later I can close my eyes and fondly smell the provolone and pepperoni.

But habituation to odors has been shown in rats to also take place within the brain, not just at the sensor level. Evidence has been accumulating in support of the hypothesis that pink noise, characteristic of complex networks, arising as it does in both single and in large collections of neurons, is the common element that connects both these observation. It has been observed that the complex neural spectrum, suppresses simple signals transmitted to the brain and inhibits simple signals, transferred within the brain. Let us consider a few examples of how the phenomenon of habituation is related to the transfer of information, using the new principle.

Consider the case of street noise coming in through the motel room window, as you lay in bed at night. The noise is typically an uncorrelated random process with the rumble of trucks and buses intermixed with sirens of police and ambulances, and the unrelenting sound of car tires on asphalt. Most people habituate to this noise, meaning that the neurons no longer fire in response to this excitation, and they fall asleep. A native city dweller may have the opposite experience and find it difficult to sleep without these reassuring sounds in the background. The generalization of linear response theory, on which the cross-correlation cube in Fig. (**5.4**) is based, suggests that the response of the brain to the street noise fades as an inverse power law, asymptotically approaching the no response zone in region II [109].

The lower plateau in Fig. (**5.4**) was described previously and the two kinds of fluctuations in this case refer to the highway noise (stimulus) and the brain activity (response). The dynamics of the brain and the external stimulus are seen to be statistically independent of one another, after a sufficiently long time. Brain activity in the relaxed non-ergodic regime is asymptotically unresponsive to ergodic fluctuation and/or simple stimuli; the complex network of neurons essentially swallows up simple signals through its complex dynamic interactions and the response fades. The suppression of periodic stimulation by a complex network, using linear response theory, was previously demonstrated by a number of investigators, who thought it indicated the unsuitability of the theory for describing the response to such stimuli. It was expected, that because the amplitude of the stimulus remains constant in time that the response should do the same. When the response was observed to fade away, linear response theory was thought to be 'dead', or less dramatically, it was thought to be inapplicable in describing the response of complex networks to simple stimuli. However further analysis [70] found that the death of linear response was grossly exaggerated and the fading of the response to the simple signal was a consequence of the network's complexity, not an intrinsic limitation of the generalized theory.

Consequently, depending on whether we hear, taste, touch or smell the stimulus, that is, the mode of stimulation determines the nature of the network response and consequently the form of habituation. Recent analysis of the coincidences between abrupt quakes in different EEG channels has led us [111] to conclude that the resting state of the brain is characterized by a power-law index of two. This value of the complexity index defines the boundary between the region of complete habituation and the domain where a response to periodic stimuli, although weak, is possible. This value of the power-law index is at the center point of the cross-correlation cube where the singular effect of pink (1/f) noise is most evident. The brain in the different states of wakefulness and sleep has different power-law index values and consequently it may also have different modes of response to external stimuli.

One way a professor can distinguish between the brain states of his students is by looking at faces and seeing what fraction of the class has light in their eyes. When I was a professor I sometimes lost the students' attention, at which time a little voice would tell me to change gears, perhaps even to ask the class where I had left them. A good teacher knows that a monotone lecture by a statue-like speaker is boring, regardless of the lecture content, because of the lack of variability in the presentation. Such lectures invariably put students to sleep and the course finds itself in region II of the cross-correlation cube.

Experience and now theory has shown that the brain processes data most efficiently when data variability is compatible with the complexity of the brain. Consequently, good teaching is very much like good acting, with ups and downs in pitch and volume, changes in facial expressions, and coordinated body movements. This variability is not just for reciting poetry, or quoting Shakespeare, but for explaining the invisible pressure on the students' skin due to being submerged in an ocean of air. Such histrionics contribute to what is being conveyed by the speaker and assists in keeping the listeners engaged. When the audience is attentive, information can be absorbed and the course jumps from region II to III.

The Principle of Complexity Management implies that information flow is maximum, when the lecturer is able to stimulate the audience, with a level of complexity comparable to the listener's brain. This does not just mean the intellectual content of the lecture, but all that goes with it. Or stated the other way around, we most readily absorb (manage) information, when the complexity of that information is comparable to the complexity of cognitive processing. In this regard our cognitive map acts like a filter, letting through what matches and blocking, or ultimately suppressing, what does not. Consequently, the more complex the cognitive map, the fewer restrictions there are on what we allow ourselves to understand.

5.2.2. Leaky Faucets and Classical Music

The sound of crashing water droplets from a dripping faucet can set a person's teeth on edge and lead to their tossing and turning at night without falling asleep. Why can't they block out the sounds of the splashing water in the way they habituate to traffic noise and the smell of strong cheese? The reason these particular sounds keep people awake lies with the spectrum of the time interval between the events, which in this case are the falling drops. In the decade of the 1970s many experiments were done to determine the relation between everyday phenomena and the then newly emerging field of nonlinear dynamics (chaos theory). One such experiment concerned determining the time interval between water drops from a leaky faucet. A laser beam, intercepting the trajectory of the falling water drops, was used to register the passage of the drop. The time is automatically recorded and the time intervals between successive splashes determined. These experiments determined that, for certain settings of the spigot, the water drops from a leaky faucet have an inverse power- law distribution of time intervals between splashes. The power-law index is in the interval between one and two. Therefore a pink spectrum describes the dynamics of a leaky faucet.

The cross-correlation model of the brain's response to this stimulus ramps up from zero to one as its power-law index increases from one to two as shown in Fig. (**5.4**). Over the brain's index interval 2 to 3 the response to the intermittent splashes is maximum as shown by plateau III in the cross-correlation cube in Fig. (**5.4**). This upper plateau indicates the parameter domain, where the brain has relatively short-memory and consequently records the sound of every intermittent drop of water. The maximum amount of information is being transferred, because the brain is not able to predict when the next drop will fall and consequently each thunderous drop is a new piece of information. This new information does not tell us of the encroachment of an enemy or eminent attack by a hostile animal, as might have resulted in its evolutionary selection. The lack of connection to present day events, urged some to conjecture that this fixation on the unpredictable, might be maladaptive in modern society. However driving in city traffic or investing in the stock market may suggest reevaluating that position.

It was empirically determined that the intermittent sequence of water drops being released by a leaky faucet is described by a Lévy stable distribution, which asymptotically is an inverse power law with an index of 1.73. Consequently, by comparing indices the power spectral density of the stimulus is seen to have a spectral index of 1.27, very close to the pink noise value of 1.0. Note that the value of 1.0 is the prediction of Wiener's Rule for the maximum response of the brain to the stimulus. This interpretation is consistent with recent experiments showing that the response of certain neurons to weak signals is enhanced by adding noise to the input. Nozaki *et al.* [112] introduced a variety of noise spectra into rat sensory neurons and determined that the enhanced response to a weak signal, through the addition of pink noise can exceed that of adding either white noise or brown noise.

Of course, it is not only annoying stimuli that fixate in our brain and refuse to fade away. As we discussed, classical music manifests a pink (1/f) spectrum and consequently resonates with the functioning of the human brain, leaving strains of melody running through a person's head long after the music stops. The persistence of music's pleasant stimulation also resides on the upper plateau of Fig. (**5.4**). It therefore appears that the fixation on the pleasantly complex and the offensively complex are two sides of the same coin.

The Principle of Complexity Matching applies whenever two complex networks interact without regard to value judgments. Various forms of the principle are useful in attempting to answer such questions as: Which person in an organization is most apt to get new ideas implemented? What can one person do to improve the morale of an organization? Why are groups resistant to suggested changes made by an individual?

5.2.3. Global Warming

Let us move to a global application of our nascent principle, getting out of the cerebral and into the geophysical realm of climate change. The notion that the recent rise in the Earth's average temperature can have catastrophic consequences is based on models that are much more elaborate than those discussed in this book. However, extended elaboration does not always guarantee deeper understanding. In science it is often the case that truly complex phenomena are only understood through a sequence of relatively simple models; often called 'back of the envelope calculations'. The name is a consequence of the literal use of an envelope or napkin on which the calculation is typically carried out, often during a meal at some restaurant over a glass of beer or wine.

For example, suppose I wanted to know the speed with which an object hits the ground, after being dropped from a 32 foot high ladder. Assume that I do not remember the solution to Newton's equations of motion for a falling body in a gravitational field. However, I do remember that the constant acceleration of gravity is 32 ft/sec^2. A back of the envelop calculation can be done using the fact that acceleration has the units of distance over time squared to determine that the object would take approximately one second to reach the ground. Consequently the speed at impact would be approximately 32 ft/sec. This estimate of the speed would, of course, be wrong, because it neglects a factor of two in the exact equation and consequently the time to reach the ground is nearly one and one-half seconds. The exact speed of impact would be 45.25 ft/sec. However it is sometimes better to be a little wrong with some knowledge, than to be absolutely right with no knowledge. That is the strength of such inexact calculations, or more accurately an exact calculation using an inexact model.

Pointing out the inexact nature of such a quick calculation is not intended to denigrate it. Scientists know well that all calculations, no matter how precise, are wrong. It is a matter of degree, not a difference in kind. In fact, the more elaborate the model, the greater the opportunity for error. This is the error resulting from the model not being a faithful representation of the phenomenon being studied, not that due to experimental inexactitude. Consider the child's swing, previously used as an example of a linear dynamic process, when it is not pushed too hard. When the seat approaches a maximum height on a level with the horizontal bar supporting the chains of the swing, nonlinear effects become important. As any adventurous child knows, the chains supporting the seat go slack and the seat drops resulting in a snapping motion, when the chain again become taught. This unstable behavior, during the return of the seat from the apogee to the perigee, has no linear description.

The global debate over climate change has two main camps, both of which agree that the average global temperature has increased by nearly a degree over the twentieth century. One camp maintains that this increase is primarily due to anthropogenic (human) causes, in particular greenhouse gases, and this group appears to have more members than the second, which is quite remarkable given the limited number of scientists that have any expertise in climatology. An apparently smaller, but certainly more scientifically conservative camp, of which I am a member, maintains that the temperature increase may be due in whole, or in part, to natural causes such as variability in the total solar radiation reaching the earth. Aside from the politics of this often acrimonious debate, I believe that the science of climate change is not yet sufficiently well developed to make firm unqualified predictions.

Of course I have reasons to hold a position that is at odds with the apparent majority opinion. One reason is that I would rather rely on my own judgment concerning what I do understand, than on someone else's concerning something I do not understand. In this regard I do not understand the output of these large-scale computer codes that are used to obtain the theoretical average global temperature. Or more accurately, I do not understand how these computer calculations relate to the temperature measurements made across the Earth's surface and the measurements of the total solar irradiance that warms the planet.

Of course I do not believe that anyone else understands them either, at least not without the auxiliary input of simple energy balance models. So let me show you what I do understand.

My colleagues and I compared the statistics of the fluctuations in the number of solar flares from 1978 to 2003, with the statistics of the anomalies in the Earth's average temperature from 1856 to 2003 [51]. In this way we treat the Sun and the Earth as two complex networks, whose dynamics we do not necessarily understand, but whose statistical fluctuations have been measured. The fluctuations in the number of solar flares and in the temperature anomaly time series, have been shown by us to have inverse power-law statistical distributions, with essentially identical scaling indices as indicated in Fig. (**5.5**). Two techniques are used to analyze the data sets for the solar and terrestrial phenomena to determine the scaling indices.

The first method of analysis is standard deviation analysis of time series. Recall that in classical diffusion the variance, which is the square of the standard deviation, in the displacement of the milk is proportional to time. The standard deviations for the solar flare and temperature anomaly time series are found to also increase as power laws in time with the power-law indices H being somewhat less than one in both cases. On the graph paper used in Fig. (**5.5**) such a power law appears as a straight line with a positive slope of value H given by 0.94 for solar irradiance, using the number of solar flares as surrogate data, and 0.95 for average global temperature time series. Note the near equality of the two values of H.

The second method of analysis is the diffusion entropy analysis in which the entropy increase of the time series from a reference state is time-dependent, yielding a new scaling index δ. On log-linear graph paper such a power law appears as a straight line with a positive slope of value δ given by 0.88 for the solar irradiance and 0.89 for the global temperature time series. The two scaling parameters H and δ have been shown to be related to one another when the underlying process is describable by a drunken Lévy walker [51]. This latter model is very much like the usual drunkard's walk, with the difference being that in this case the drunk has very long legs and can therefore intermittently take larger steps than usual.

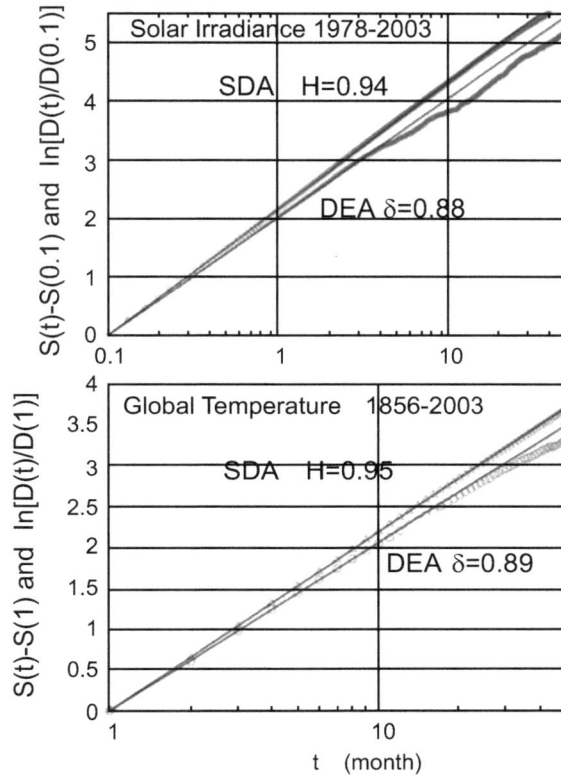

Fig. (5.5). The diffusion entropy analysis (DEA) and the standard deviation analysis (SDA) are used to analyze time series from both the Sun and the Earth time series data. The analysis of the ACRIM composite total solar irradiance time series (1978-2003) in the top graph yields the two straight lines with scaling parameters indicated. The analysis of the global temperature anomalies time series (1856-2003) in the lower graph yields the two straight lines with scaling parameters indicated. [Adapted from [51]].

If the underlying statistical process is Normal then the two scaling parameters H and δ would be equal to one another. However in the general case, when the statistics are not Normal, there is no reason for these two parameters to be related, they are typically independent scaling parameters. In the drunken Lévy walker model the two parameters are related in the manner indicated by the solid curve in Fig. (**5.6**) for both parameters, between the values 1/2 and 1. The theoretical relation between the two scaling parameters is satisfied remarkably well by their empirical values as depicted in Fig. (**5.5**) [51] and very near the theoretical curve as shown by the dot in Fig. (**5.6**). When theory and observation agree this well in a nontrivial comparison, it is considered by some scientists to be compelling evidence that the theory is at least partially true.

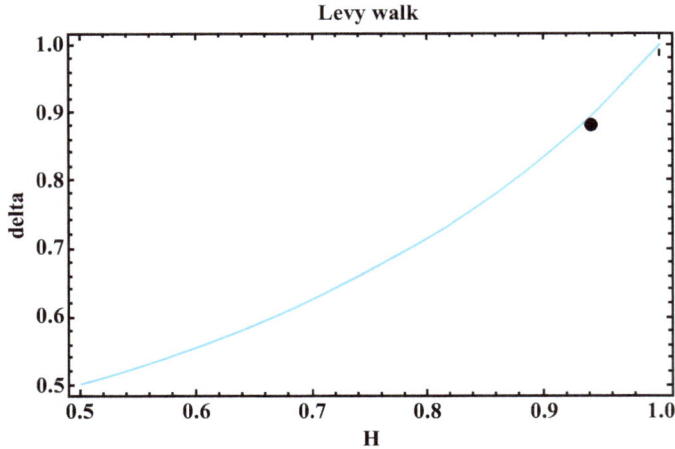

Fig. (5.6). The relation between the two scaling parameters H and δ is determined by a Lévy drunkard's walk as indicated by the solid curve. The empirical values of the parameters from both the solar flare data and the Earth temperature anomaly data overlap in the indicated dot.

My naive explanation of the fact that the fluctuations in the radiation from the Sun and the fluctuations in the Earth's average temperature are virtually identical is based on an information transfer perspective using the cross-correlation cube. Here we argue that the information exchange is nearly resonant, being only slightly off the transition point from the no-influence lower plateau in the cube, to maximum influence on the upper plateau in Fig. (**5.4**). This sudden jump occurs at the point where both the stimulus (Sun) and response (Earth) have pink noise variability. The Earth's unperturbed power-law index is suppressed, using Wiener's Rule, and subsequently replaced by the solar flare power-law index, indicating that the statistical fluctuations in the Earth's temperature field are dominated by the fluctuations in the total solar irradiation. Traditional thermodynamic arguments based on the size of the energy transfer would maintain that the influence of the Sun on the Earth's temperature change is negligibly small [51] and such arguments have been put forward. However, from the present perspective of understanding complex phenomena, using simple models, the cross-correlation cube indicates that the change in the Earth's temperature is an information-dominated process not an energy-dominated one.

Finally, I take finding that the two scaling indices H and δ satisfy a theoretical relation, as evidence for the strong influence the variability of the total solar irradiance has on climate change. I have every expectation that the arguments in this section will be dismissed as simplistic by some and as plain stupid by others, but as the science unfolds over the next decade or so; after the shouting has died down and the personal abuse fades; it should become apparent who is the prince and who is the frog.

5.3. PINK (1/F) NOISE

Pink noise appears, as reviewed by West *et al.* [71], in the movements of the human body, such as in the time interval of strides during walking, in the two-dimensional motion of postural sway, and movement in synchrony with external stimulation, such as in marching to the soothing voice of a Drill Instructor counting cadence; also in physiologic networks as observed in heart rate variability, human vision, the dynamics of neuronal dischages within the human brain and in cognition; finally pink noise is measured at the level of single neuron adaptation to various stimuli. The original assertion that the power-law index in the pink noise spectrum is 1.0 was subsequently shown to extend the spectral index to the broader range from 0.50 to 1.5, see West *et al.* [71] for those that want to dig a little deeper. As they point out, the variety of mathematical models used to explain the generation of pink noise is as wide ranging as the disciplines in which scaling phenomena have been observed.

The American mathematician George David Birkhoff (1884-1944) was interested in how we humans make aesthetic judgments. He formulated a theory of aesthetic value that postulated that the most appealing artwork has sufficient variability to maintain the viewer's (listener's) interest; art that is too regular becomes boring. On the other hand this lack of predictability should not be overwhelming; there should be enough underlying structure to guide the audience to the next surprise. Here again the tension between prediction and surprise is present with the most appealing structure arising from a balance between the two. This is completely consistent with the findings of Voss and Clarke discussed earlier, where people judged the tightly correlated spectrum of brown music to be boring and the structureless hiss of white music to be irritating. The aesthetic balance is, of course, provided by pink music, with its spectrum, that diminishes as an inverse power law in frequency, with an inverse power-law index near one.

In the explanation of why so many phenomena have inverse power-law statistics we relied on the scale-free property of the logarithm. In this way we determined that an inverse power law characterizes the statistics of a network, with maximum disorder, consistent with scale-free behavior. A waiting-time distribution characterizes the probability of waiting a given time before an event occurs after the last event, can be derived in this way and leads to the concept of fractal time as extensively discussed in a social context by Vorbel [113]. The spectrum corresponding to fractal time is typically pink, which is to say, it is 1/f noise.

The time registered by the ticking clock is vastly different from the human experience of time. A person's experience of a time interval is determined by what they are doing. That experience was eloquently captured by Thomas Mann [114]:

> "Emptiness and monotony may dilate the moment and the hour and make them 'tedious'; the great and greatest periods of time, though, they shorten and fade away even to nothingness. Conversely, rich and interesting content is capable of shortening and quickening the hour and even the actual day; on a large scale, though, it endows the course of time with breadth, weight and solidity, so that eventful years pass much more slowly than those poor, empty, light years which the wind blows before it, and which fly away. So, actually, what we call tedium is, rather, a pathological diversion of time, resulting from monotony: in conditions of uninterrupted uniformity, great periods of time shrivel up in a manner which terrifies the heart to death....."

The long periods of tedium disrupted by short intervals of intense activity, followed by even longer intervals of interesting bustle, interspersed with pulses of ennui are typical of intermittent experiential time intervals. The key word here is experiential, it is what a brain does with the time given to it by the clock to process activity. This particular chestnut, regarding clock time and mental time, has been roasting, since the middle of the nineteenth century and it is only now that we are able to quantify the distinction between the two. For example, the influence the statistics of mental time plays in the phenomenon of forgetting can now be quantified.

5.3.1. Memory

The first systematic laboratory studies examining the properties of human memory were published in 1885 by Herman Ebbinghaus [115]. He maintained that repetition strengthens the association made by committing something to memory. He tested his theory by committing long lists of nonsense symbols to memory for various numbers of repetitions. From these experiments he determined that learning increases more rapidly at first and then more slowly with the number of repetitions, resulting in the first learning curve. Similarly, he recalled given lists of symbols, after varying lengths of time subsequent to learning and interpreted the response as constituting the degree of memory. As with memory, forgetting increased more rapidly at first and then more slowly, leading to the first forgetting curve. The functional form for these curves had been incorrectly assumed by many to be exponential [116].

In the experiments on himself, Ebbinghaus measured how much memory is retained after the cessation of learning. As we [117] explained elsewhere, he did this by measuring the amount of relearning required, after a given time, and found a functional relation between the memory saved, that is, the memory retained, and the retention time. If the functional form of the learning and retention curves were in fact exponential, the simplest dynamic description would be a psychophysical rate equation for the strength of the performance measure. In other words, it would be like the Malthus equation for population growth. However, as shown by Ebbinghaus and a number of subsequent investigators the decay of memory is not exponential, as is verified in Fig. (**5.7**) using the Ebbinghaus data for retention of memory. This figure is a log-log graph, whose vertical axis records the data for the memory saving function of Ebbinghaus and the horizontal axis denotes the time in days since the cessation of learning, which we have replotted [117]. The slope of the line segment in the figure is interpreted as the rate of forgetting, the greater the slope the more readily we forget.

We [117] hypothesized that pink noise is the common element that may explain both the loss of memory with time and the change in the rate of forgetting with time. We tested this hypothesis using a theory of learning that involves chunking and by introducing a discrete theory of forgetting called chipping. A chunk of memory is an aggregate of elements that have strong associations (correlations) with one another, but weak correlations with elements within other chunks. Note that this definition of a chunk bears a striking similarity to that of the small-world theory of complex networks [102]. Moreover, Zemanova *et al.* [118] point out that the anatomical connectivity within the brains of lower primates display features of small-world and scale-free networks. We [117] modeled forgetting as a process of chipping away at a chunk of memory and one or more elements of memory are lost with each chip. The chipping process is characterized by a random sequence of discrete events, each event being a chip that disaggregates, breaks loose, an element of memory.

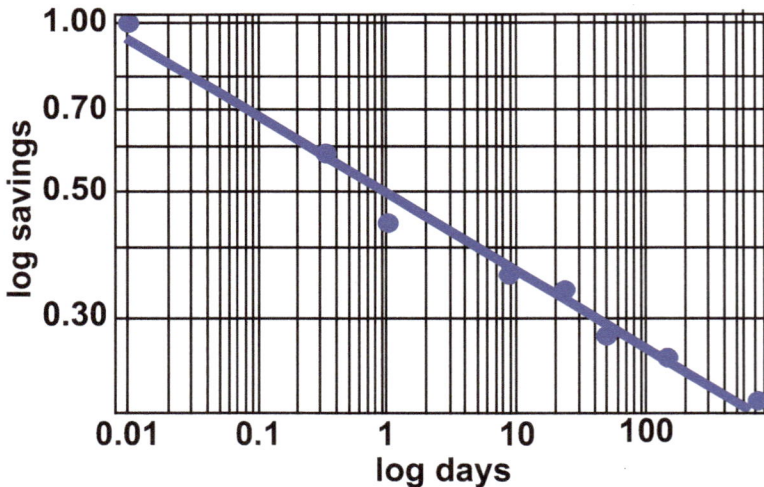

Fig. (5.7). The closed circles are the data from [114] for the memory saving function and the line segment is the best mean-square fit to that data. The data are clearly well fit by an inverse power-law function (solid line segment) with a slope of -0.14. [Adapted from [117] with permission].

A more complete theory of memory and forgetting would include an understanding of the interrelations between mental processes starting from neural networks, to single neurons, down to individual molecules. The chipping model does not do this, but instead it focuses on the psychophysical aspect of memory and forgetting, both of which are ultimately embodied at the level of proteins and molecules. It has recently been determined that certain molecules (tyrosine phosphatase corkscrew) augment memory formulation with spaced resting intervals and is called the 'spacing effect' and appears, for example, in how long common fruit flies (Drosophila) must rest between learning sessions to form long-term memory [119]. During these intervals the 'corkscrew' molecules initiate significant change in chemical activity; activity that is critical for communication between cells. The formation of long-term memory only occurs when the resting interval is sufficiently extended to allow the chemical process to reach completion. The occurrence and function of these chemical processes have been established, but the details of this molecular mechanism remains to be investigated. It has also been found that a mutant form of the corkscrew molecule prolongs the time of the resting intervals before the formation of long- term memory [119]. Such mutations induce forgetting at the molecular level. It might be an intellectual stretch, but it is worth considering that the spacing effect is quite possibly related to the observation made by Ebbinghaus [115]. The observation that forgetting is inhibited by distributing training over several sessions, rather than compacting an equivalent amount of training into a single session. Distributed learning is consistent with the chipping model of forgetting. Consequently, effective learning is not a marathon, but is a sequence of sprints separated by intermittent periods of rest.

5.3.2. Decision Making

Another problem area Professor Grigollini and his students studied and in which they allowed me to participate was that of decision making. When a scientist ventures beyond his/her area of expertise, it is wise to study what others have previously attempted and accomplished. In our review of the literature on decision making, we found that the history of probability parallels the history of how we make choices, with incomplete information and under conditions of uncertainty. Consequently, it is not surprising that the mathematical formalism of probability theory was initiated in gambling. Choice enters gambling in the determination and selection of the most likely outcomes in games of chance. The players on both sides of a wager want to win and therefore in the initial efforts at formulating a general theory of probability, a great deal of attention was paid to developing measures of play that could be used to maximize winnings, or at least minimize losses [117].

It has been noticed by a number of investigators that in behavioral-decision research, people often fail to choose what will make them happy, either because they do not know what that is, or they fail to act on what they do know, see [117] for references. Ariely, in his remarkable book *Predictably Irrational* [120], takes the position that undesirable outcomes can be predicted and avoided and he explains the fundamentally irrational nature of many decisions. It is not that psychology dismisses rational thought, as Wittmann and Paulus [121, 122] point out. Instead they emphasize that other factors, such as satisfaction and impulsive behavior become of equal, or more, importance in the psychology of decision making. Regret is one of the most studied emotions that mitigate rational decision making , and regret aversion often dominates the decision-making process. On the other hand, curiosity, which can be associated with impulsive behavior, can suppress or kill regret. Sociology acts on a broader stage to examine how peer pressure and leadership often dominate decision making in human networks.

Experiment supports the intuitive position that decisions are often tied to rewards. The time between making a decision and its subsequent reward plays a significant role in the decision-making process. This delay-dependence of a decision is only part of the story, however. It is also reasonable to expect that the influence of time between decision and reward is mitigated by the magnitude of the reward. In a rational world the longer one must wait for a reward, the smaller is its experienced value at the time of the decision. Note that this is not the same as the experience of the reward after the elapsed time. Samuelson proposed the functional form for the value experienced (utility function) by the individual making the decision in 1937 to be an exponential function of the delay time. There was no fundamental economic or psychological reason for Samuelson to make this choice except that, like Malthus, it was the simplest mathematical function that satisfied his mental map. But, as we have repeatedly found, the world does not always conform to our notions of consistency.

The monotonic decay of the exponential with delay time is consistent with the accepted notion of rationality, in which the longer the delay time, the greater the discounting of the reward. Another property of the exponential, that we have encountered before, is that it can be characterized by a single parameter, the empirical rate of decay; the utility is discounted at the same rate in each interval of time. Unfortunately, the studies in behavioral economic violate the economist's assumed consistency in inter-temporal choice, but rather there is an inconsistency in human inter-temporal choice and this is rightly interpreted as an implied irrationality. The consistent inter-temporal choice predicted is not what is observed experimentally.

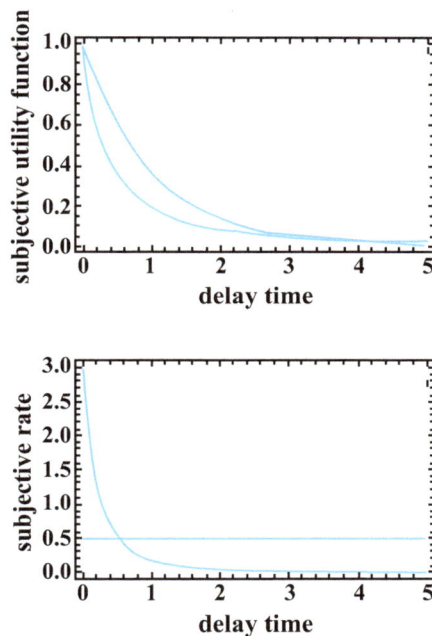

Fig. (5.8). On the top are graphed the exponential and generalized hyperbolic distributions with the exponential decreasing more slowly. On the bottom are graphed the rates of the same two distributions. the exponential has a constant rate (flat line) whereas the hyperbolic distribution has a decreasing rate over time.

The kinds of distribution discussed are plotted in Fig. (**5.8**). On the top panel the exponential and generalized hyperbolic distribution are graphed for selected values of their parameters. The hyperbolic distribution is seen to plunge more rapidly than does the exponential at small values of the variable. However at some point in time the tail of the hyperbolic distribution takes over and the hyperbolic crosses the exponential. In the lower panel the rate of decrease of the two distributions are depicted. The exponential has a constant rate of decease, but the hyperbolic has a rate of decrease that becomes smaller with time. Eventually the rate of decrease of the hyperbolic distribution crosses the rate of the exponential. If the exponential with its constant rate depicts rationality, then a rate that decreases in time with no new information must denote irrationality.

As discussed in connection with the applications of the Normal distribution, psychophysics was invented by Fechner [123] to assign objective physical measures to experiential phenomena in psychology. The history of experiments include individual response sensitivity to changes in light, heat, electric shocks and other such physical stimulations. More subtle experiences have been addressed experimentally, such as the perception of time and the pursuit of happiness, and the failure to act on a prediction that insures success is related to rationality. Our own research [117] was concerned with the relation between the perception of time by the human brain and the time as measured by a clock. Moreover, how the difference between the two influences decision-making processes.

The notion of delay naturally enters into the description of complex random phenomena, describing the time interval experienced, between when a decision is made and when it is rewarded. This perspective requires the subjective time to be a random experiential variable, rather than an objective parameter determined by a clock. The delay probability, as characterized by the subjective utility function, was constructed [117] using Fechner's law . This law, in the present case, relates the subjectively experienced magnitude of mental time to the objectively measured magnitude of clock time. The associated subjective utility function is the probability that a decision's reward has not been received up to a given time. Fechner's law transforms the rational (dynamically consistent) subjective time into the irrational (dynamically inconsistent) time given by the Pareto distribution [117].

A generalized hyperbolic form of the utility function, which is a generalization of the Pareto distribution, was postulated by Takahashi [124], with parameters he determined experimentally. In one set of experiments individuals made decisions regarding their own utility. In a second set of experiments individuals made estimates of the utility of the decisions made of others. From these experiments Takahashi deduced two sets of parameter values, which were found to be quite different. Our own calculations [117] independently gave an excellent fit of these data to the stochastic utility model.

In the present psychophysical model the power-law index provides a measure of how soon individuals change their minds with the passage of time, but without new information. The change in the rate of decay over time quantifies irrationality in terms of the deviation of the measured distribution from the exponential utility function of Samuelson. The deviations from the exponential, using Takahashi's data, indicate that irrationality in intertemporal choice is stronger when the outcome is irrelevant to the decision-maker, with smaller values of the power-law index corresponding to greater irrationality. It is well known that such generalized hyperbolic (asymptotically inverse power law) delay-time distributions have diverging mean times for a power-law index less than two. This is the observed situation for decision making.

The exponential form of the discounted-utility model proposed by Samuelson rests on a constant rate, as mentioned earlier. Moreover, even if we assume ordinary people are internally rational, as done in this decision making model, the decisions they make in the external world are characterized by the asymptotic inverse power-law utility function and therefore appear to the world to be irrational. The concept of time is subjective so that a strongly impulsive patient has an experience of time that is very different from that of a control individual. Stimulant-dependent individuals estimate that a given time interval is longer than the corresponding estimates made by control subjects [122]. Consequently, postponing gratification has a telescoping effects on the discomfort level of stimulant-dependent individuals.

In Takahashi's experiments the impulsivity, as measured by the response time in the stochastic inter-temporal choice model, is a factor of ten shorter when the outcome is irrelevant to the decision-maker. The experimental subject estimates that unknown others make choices ten times faster than their own reasoned decisions. The interpretation of this latter observation is interesting, since a given individual estimates others to be less rational and more impulsive than they themselves are, leading to a number of speculations. For example, lawmakers may, for a number of reasons, unconsciously assume they know what is best for their constituency even without asking them. Thus, it is also important to emphasize that the response to these stimuli change according to whether the brain is that of a healthy person, or one that is stimulant-dependent.

5.4. NETWORK-CENTRIC WARFARE

In adapting to our information-rich networked society the Department of Defense reevaluated the historical notions of war and created a new theory called Network-Centric Warfare [125]. This new theory has at its core a shift in focus from platforms such as trucks, tanks, and planes to a networked Army of sensor webs, computer networks and intelligence gathering social web-works. The implication of this shift is that members of the military are no longer viewed as independent individuals, but are part of a continuously adapting complex ecosystem, both friend and foe alike. The military is invariably a reflection of the society of which it is a part, so this adaptation to networking is a consequence of basic changes in western social interaction over the past two centuries. These changes refer to fundamentally important strategic choices civilians have made in order to adapt and survive in a changing social ecosystem.

Taken individually these social changes often appear trivial. For example, the art of letter writing has been all but lost, because we now twitter, tweet and email. At the turn of the nineteenth century human contact over long distances was made by means of letters, that required days or even months for delivery. Consequently letter writing was considered an art form by many people and hours were spent in their composition. But that century saw the introduction of the telegraph, a boon for business that drastically reduced the time spent in transferring vital information. However, the personal connection of the letter was still safe. The twentieth century witnessed the telegraph's being replaced by the telephone and the use of the human voice for contact. The time scale for communication from one coast of the United States to the other decreased from tens of days to perhaps minutes, a time contraction on the order of tens of thousand. The twenty-first century with the wide use of computers, Skype and iphones brings in another factor of one hundred or so in the reduction in information transfer time. I estimate the change in time scales from delivering letters to exchanging emails to be on the order of one hundred thousand or the time difference between reading this book and blinking.

Letters and emails are representative of the fundamental differences between the industrial age and the information age; the era of manufacturing and that of the bit and bite. In the former a material object was passed between individuals to denote the intended transfer of wealth or power, in the latter case a brief electronic message serves the same purpose; the difference between a letter of credit and a computer transfer of funds. Consequently the distance separating businesses and individuals has lost the importance it once had in human affairs and the significant time interval is no longer that for transferring information, precisely because it is thousands of times shorter than it once was. It is the time to compose the message that limits the speed of communication and the time to transmit the message can, for most purposes, be ignored. The tempo of modern living is thousands of times faster in the information age, because there is no mitigating influence to lengthen the time between decisions. Technology has reduced time to that required to think up the input for the next communication. There is no waiting time in which to consider alternatives, or to devise a plan B, there is what you know and the need to make a decision. Now!

The flash mob and flash crash flow from the compression of space and time in the information age. Those that make the best use of the technology have an information advantage over their competitors, and unlike the static notion of knowledge that existed when the expression "Knowledge is power" was first coined, knowledge and information are now dynamic quantities and we have the Red Queen effect. This effect is taken from *Alice in Wonderland* where after running with the Red Queen for a long time Alice noticed they had not moved and said how different that was from her country:

> "A slow sort of country! said the Queen. Now, here, you see, it takes all the running you can do to keep in the same place. If you want to get somewhere else, you must run at least twice as fast as that!"

In the same way the operations tempo of the modern Army has been increased to match that of society. The tempo is intended to minimize the miss-match in times, between the interaction of the Army with the world at large. But more importantly, this tempo now permeates upward from the newest recruit to the General Officer, no one can escape the domination of the emails and the networking they represent.

The military strives to adopt strategies that work and those strategies are extracted from the societies to which they belong. Scientists explain why the adopted strategies work; or as is more often the case, they apologize for why they do not work in the military in quite the same way as they do on society's broader stage. Recall that an apology is a formal, after-the-fact, justification for why things turned out the way they did. Scientists have rarely been in the position of being able to predict what will work in the human sciences and what will not. In the present context such uncertainty has to do with the dominate kinds of complexity being investigated in the two social contexts. The sociophysics of the nineteenth century recognized this and acknowledged the importance of statistics. Unfortunately, the scientists of that day limited themselves by accepting the law of error as an explanation of observed, but not understood, variability. A perspective that is totally inadequate for the information age.

Today scientists are attempting to face complexity head on and consequently information itself is seen to be uncertain, only partly relevant and to have a limited shelf-life. An organization, with information superiority, has that superiority for a short time during which it must decide what to do, as well as, to implement that decision. This is as true in society as it is in the Army. In asymmetric warfare (Information Warfare), information superiority ebbs and flows between friend and foe, and those that anticipate what to do at the next stage of superiority certainly gain short-term advantage. Those that learn how to systematically use information superiority at one time, increase the likelihood of its occurring at a later time, and thereby gain long-term advantage. This is the stitching together of logistic curves seen earlier, with regard to technology replacement, see Fig. (**1.13**).

In asymmetric warfare the superior force does not always win, because the battle space includes all of society and not just the battlefield. The battlespace is that of the information age and the differences in the networking of the Army and that in the insurgent force becomes all important. The rules of each are different, as is their intent, so who the enemy is and what their purpose is, are often not apparent. A western society may build an army, whereas a middle eastern society may infiltrate a school. In this context the distinction between combatant and civilian is blurred to the point of indistinguishability. All this and more is captured in the bumper sticker: "Winning the war and losing the peace".

Today's theatre of war is the downtown district of any city; a district shielding combatants camouflaged in civilian clothes. Urban warfare technology is not that of traditional planes, tanks and bombs, but consist of cyber attacks and remotely detonated improvised explosive devices (IEDs). In modern warfare the government of the enemy often does not speak for the people, so it is not politically acceptable to bomb a city taken over by the enemy. The warfighter must determine and destroy only those buildings in which the true enemy resides. Command deploys soldiers into an urban environment who are trained to selectively kill only certain members of the indigenous population. The enemy is not on one side of a bright line with friends on the other; the two are mixed together like the fingering of oil in water. The enemy is both far away and in the house next door. This is part of the new complexity of warfare that is being addressed in the twenty-first century and requires a new kind of situational awareness on the part of the soldier. It is not the individual that is being identified, but rather it is the connection of the individual to a social network that gives the soldier the certainty to shoot and kill. This is one of those deep lessons that is part of Network-Centric Warfare.

Another part of the complexity of the new kind of war has to do with the missions a soldier is required to carry out, in addition to traditional combat. These activities change with the changing mission of the Army, from domestic disaster relief to global conventional war. What does not change is the necessity for the military to be well trained across the complete spectrum of military operations. The training required to prepare the individual to support domestic disaster relief, to participate in counter-terrorism efforts and to carry out limited conventional conflict, is dramatically different from what it was in the past. One aspect of training is that for the specialist, another requires the flexibility that enables a soldier to give humanitarian assistance, be a peace keeper, quell civil disturbances and yet be ready to enter into limited conventional conflict should the situation change. The necessity to understand the social dimension of the enemy, such as the stability and sensitivity of social networks, like those of the terrorists, has never been greater.

The detailed arguments regarding the fundamental importance of Network-Centric Warfare are logically consistent and intuitively attractive [125], but they are too voluminous to be repeated here. Instead we identify the four foundational tenets of Network-Centric Warfare and suggest how they might be supported by the Principle of Complexity Management. These four tenets constitute a working hypothesis [125]:

1. A robustly networked force improves information sharing.
2. Information sharing enhances the quality of information and shared situational awareness.
3. Shared situational awareness enables collaboration and self-synchronization, and enhances sustainability and speed of command.
4. These, in turn, dramatically increase mission effectiveness.

I should point out that the tenets have neither a basis in controlled experiment, nor scientific theory. An attempt at providing such a logical foundation was given by Garstka and Alberts [125], but that analysis took the form of case studies, not controlled experiments. Consequently the discussion did not rise to the level of systematic laboratory or field observations capable of prediction, or of testable predictions. Alberts *et al.* [126] emphasize that Network-Centric Warfare is about people, organizations and processes, but it is also about the hardware that enables information flow, through the interconnections of these networks. They go on to prescribe how one could experimentally test the tenets and are clear on the fact that such experiments have yet to be done.

The cross-correlation cube is the first formal device that can be used to address this information flow and assess its consequences on Network-Centric Warfare.

The limitations of the first tenet can, in part, be determined by the Principle of Complexity Management and the degree of complexity of the networks across which information is being shared. The matching of the power-law indices shown by the cross-correlation cube determines the degree of situational awareness maintained by the networked soldier receiving the information.

Network-Centric Warfare converts information advantage into competitive advantage by creating processes and procedures through networking that otherwise could not exist and potentially leads to increased mission effectiveness. This involves the coevolution of the organization with the process, and without such coevolution the organization could be suffocated in an avalanche of data, that it would be unable to transform into information and ultimately into knowledge. To paraphrase Alberts *et al.* [126] the road to warfare based on this new way of thinking needs to be richly populated with analyses and experiment to reap the potential benefit and avoid the pitfalls of unintended consequences. Moreover Network-Centric Warfare purports to facilitate the entire spectrum of military operations, from peacekeeping, deterrence, and dissuasion to violent clashes and sustained high-intensity conflict including the countering of irregular catastrophic and disruptive threats.

The United States has been joined by the majority of NATO countries in adopting this new theory of war, under the rubrics *Network Enabled Capability* in the United Kingdom and elsewhere and *Network Based Defense* in Sweden. Consequently, decisions are made every day that affect the entire world, based on the correctness of Network-Centric Warfare; a theory without scientific confirmation, as yet.

5.5. WRAP-UP

One set of dots that may not have been connected is that, as yet, there is no Network Science. There are social networks, transportation networks, terrorist networks, communications networks, information networks, neural networks, ecology networks, *etc.* and there are multiple theories for each of these networks, as well as network theories that span multiple disciplines - but none of this constitutes a Network Science. Why? Because the various theories do not follow

246 Simplifying Complexity: Life is Uncertain, Unfair and Unequal

from a set of first principles that are independent of mechanisms in specific disciplines. As Einstein once observed: "We can't solve problems using the same kind of thinking that we used when we created them". Even the phenomena of emergence that has been largely explained using complex adaptive systems lacks a principle that can identify when a complex network will manifest a specific emergent property. Swarm logic anticipates that complex networks will have emergent behavior, but not what that behavior will be. The closest the analysts have come to this desired goal is the application of the generalization of linear response theory to complex physical and non-physical networks, giving rise to the new principle on how to manage complexity.

The Principle of Complexity Management [71] is only now undergoing the close scrutiny of the scientific community that is necessary for wide scientific acceptance. Of course, it is not the acceptance, but predictions and experimental testing of those predictions that establishes the veracity of a theory or principle. Complexity management has a firm mathematical foundation and has been useful in addressing a number of apparently disconnected phenomena from a psychophysical perspective, including habituation, fixation and the new theory of warfare. However this principle has not yet successfully predicted the existence of a previously unobserved phenomenon. By successful I mean that a predicted phenomenon has been experimentally confirmed to exist. Until that time the theory of networks will remain without principles with which to formulate a science.

CHAPTER 6

Apology for Complexity

Abstract: This chapter provides a summary of the material discussed; highlighting what is important and connecting those parts of the story that might have been obscured in presenting the details. This apology is my understanding of the formal justification for complexity in the real world. In turn, it is an examination of what complexity implies, about the difference between how we react to what we have, as opposed to reacting to what we want, but do not have. People always respond to events according to their mental maps of the world. Consequently, when they find the response to be inappropriate, the most reasonable thing to do is change the map. However, people are not always reasonable or logical.

Keywords: Chapter summaries, Complexity, Epilogue, Fractional reasoning, Highlights, Illogical, Individuality, Linear, Mental maps, Nonlinear, Organization man.

The final chapter of a book of fiction should be climactic; pulling together all the various plot lines in creative, fulfilling and even unexpected ways. When the final synthesis is done well the reader puts the book down with a feeling of satisfaction; when done by a master the reader is reluctant to put the book down at all. I maintain that a truly interesting work of non-fiction should aspire to the same end. Therefore this final chapter is a challenge for me. It is similar to retiring to the library, after a dinner with friends, for good company, cognac and, if you are so inclined, a good cigar. That good company is herein provided by a brief summary of what has been discussed in the first five chapters. The cognac, or brandy, takes the form of an introduction of a mathematical infrastructure, with which to clarify why the inverse power law is ubiquitous. The mathematics is that of the fractional calculus, in which non-integer integrals and derivatives appear, and that provides a single perspective, from which the nine different mechanisms that generate inverse power laws discussed in Chapter 4, can be understood. Of particular interest is the cigar smoke that ties the critical dynamics of complex networks to the fractional calculus; thereby reducing a many-body system to a one-dimensional rate equation.

6.1. SUMMARY OF CHAPTERS

This essay began the discussion of complexity in the first chapter, with the assertion that scientists think about complexity differently from most people. They do this in part because they are trained in method; but perhaps more importantly they do this because they are drawn towards quantitative models that enable them to think systematically about complexity, without passion or bias. It is not that scientists do not have passion or bias, for they most certainly do, but they can step beyond those limitations in their reasoning and come back to them once the reasoning is complete. Scientists, like other people, mentally store experiences that sprawl into a cognitive landscape forming a mental map of the world. This map is used in qualitative assessments and quantitative reasoning to provide an individual's unique lens, through which they see and understand the unfolding of events. The essay's narrative voice offers one scientist's point of view on how people understand the complexity of every day phenomena and how their mental maps mitigate their response choices.

The life-long process of developing mental maps begins at the earliest stages of childhood and determines whether the context in which all subsequent experience is stored is hostile or friendly, detached or engaged, coherent or confused. The formation of these maps are made explicit and analytic by youths who grow into scientists. These consciously developed maps modulate how scientists think about difficult problems and morph into quantitative models, which can be shared with like-minded individuals and whose predictions are amenable to experimental testing. The weaknesses of mental maps are revealed through systematic comparisons with real world data and these mental maps converge on reality by continuously changing them to eliminate inconsistencies. Chapter One finishes with the realization that what we think we know, but which is not true, is the highest barrier to understanding the nonlinearity and complexity of everyday experiences.

Chapter Two maintains that western civilization and the industrial revolution fostered adopting a particular map of the world; one based on the clockwork universe of Newton. Even when things become complicated and unpredictable, they do not differ too much from the mechanical world view. Uncertainty is described by Normal statistics, the mean characterizes the process and the scatter around the mean is not very large. This is the linear additive world view where fairness can be quantified and inequality can be measured. In this world industry flourishes, everyone has a job and failure can be controlled. The Utopian dreams of social order by people such as Marx, even if not a natural outgrowth of social evolution, from their perspective, can and should be imposed.

The historical reasons for the linear additive world view are sketched out and its consequences are explored. The fundamentals of uncertainty, probability and statistics are explained without burdening the reader with extraneous mathematical formalism. However, the discussion of the background concepts discloses the empirical evidence supporting the acceptance of Normalcy and traces the underlying reasons for Normal statistics, as well as, the theoretical reasons for its development in the physical sciences and its inclusion in the human sciences. This chapter explores those areas of science that have come under the spell of Normalcy and its compatibility with the great nineteenth century social thinkers.

Chapter Three examined the symmetric behavior of the Normal distribution and finds it to be inconsistent with data from such everyday phenomena as the distribution of income, the variability of prices in the stock market, and the intermittency in time intervals between emails or letters. A fundamental imbalance in the distribution of income was discovered empirically at the end of the nineteenth century by the political scientist/sociologist/engineer Vilfredo Pareto. The Pareto distribution has subsequently been found whenever the process examined is complex; from the number of earthquakes [127], to the number of sexual liaisons; from the size of a person's paycheck to the length of their email messages [59]. The real world is nonlinear, multiplicative and terrorized by extreme events, such as flash riots, crashes of the stock market, congestion of healthcare delivery systems, and power blackouts. The Pareto distribution suggests how to quantify a society's degree of unfairness using the power-law index to measure complexity.

The exploration of data and theory divulges the secret of the human sciences, that being, that Normalcy is a myth. Data from complex networks systematically discloses that all such phenomena, whether from the physical or human sciences, are described by non-Normal statistics, which have Pareto distributions with long tails. The existence of these long tails implies an imbalance in how the quantity being measured is distributed and that imbalance determines unfairness, not only in the distribution of wealth, but in the frequency of surprises as well (tipping points). It is not only the unfair allocation of stuff that the power law entails, but the tyranny of extreme events (black swans), as well as, the fear of their occurrence, that dominate our lives; the failing grade, being passed-over for promotion, losing a loved one, are experiences that shape who we are. The power-law index is proposed as the new measure of stability in the world of Pareto. This measure replaces the mean and standard deviation so long held scared in the world of the Normal.

Chapter Four presents a brief introduction of networks, necessary for understanding complexity. The coupling of elements within a network, across scales is crucial for distinguishing between the neural network, that is the human brain, the computer links that form the World Wide Web and the communications webs that support the global market. There is a finite number of mechanisms that describe the connectivity of these and many more networks, all producing the kind of distribution revealed by Pareto. As the architecture of phenomena increase in complexity, the various guises in which the inverse power laws appear increase as well.

The science underlying the appearance of complexity and what its informal implications are for complex networks is disclosed in Chapter Four. The clustering of events, interspersed with quiescent intervals of various lengths, is characteristic of complexity. Familiar phenomena are discussed in which intermittency in time determines the human response, such as why leaky faucets are so annoying and music can be so pleasant. This same bursting behavior occurs in space, as well as in time, and the heterogeneity in space explains why some people become 'stars', while most people remain 'extras' in their own lives. The examination of theory uncovers multiple mechanisms of varying degrees of granularity, that generate complexity, such as the rich getting richer, the amplification factors available to the wealthy and the domination of increasing entropy. This chapter contains the explanation as to why so much of what people do, say, and think, is determined by a Pareto distribution.

Chapter Five speculates on why scientists love to formulate laws and principles that capture what is known, for example, the conservation of energy. It is partly due to the fact that they use such principles to understand the new and exotic. The study of complex networks has produced at least one candidate for a new principle, the Principle of Complexity Management. This proposed principle shows how complex networks control one another, through the exchange of information; not unlike their physical cousins, that use the exchange of energy. Information networks may consist of social organizations, collections of interacting neurons or swarms of birds, whose complexity is quantified by pink noise and measured by the Pareto index. The Principle of Complexity Management offers a single explanation as to why we doze off in front of the television, or while reading a book, and yet a leaky faucet, or classical music, holds our attention. This nascent principle may even explain some aspects of global climate change.

It bears repeating that this new principle was discovered through the study of how information is exchanged between complex networks, such as between a book and a brain. The flow of information explained how animals habituate to strong odors and noise, however, this is only part of the principle. The cross-correlation cube captures the full domain of application of the principle and shows why we pleasantly focus on classical music and annoyingly fixate on leaky faucets. The power law of Pareto is determined to play an integral part in how information is exchanged between complex networks, using the cross-correlation cube whose center point reveals the dependence of complexity on pink or $1/f$ noise. The cube intertwines pink noise and the Principle of Complexity Management to explain human memory, decision making, situational awareness and the myriad of other responses the individual has to a complex environment.

6.2. THE INDIVIDUAL AND THE ORGANIZATION

Up to this point we have concentrated on the mental maps that individuals make of the world, but have not addressed how that map can influence the behavior of the individual. Perhaps the most insightful book addressing how the world affects the behavior of the individual was written for the generation returning home from the Pacific and from Europe at the end of the Second World War; *The Organization Man* by Whyte [128]. An organization man is someone who has so completely adopted the attitudes and behaviors expected by a company, that he or she has relinquished their individual identity. Whyte attempted to resolve two conflicting points of view regarding this loss of identity. On the one hand, he discusses the American tradition of the rugged individualist, overcoming all obstacles, using individual creativity to realize his/her dreams. On the other hand, he argued that there is safety in numbers and many of those returning from the military believed organizations make better decisions than do individuals. Consequently, working within an organization is preferable to working alone. An analysis of these two perspectives led Whyte to argue for the creativity of the individual to produce better results than those made by the anonymous risk-adverse managers of the corporation. The arguments are compelling, but whether or not one is convinced, one way or the other, often depends on one's previously held beliefs.

In a more contemporary scientific setting, the change in behavior of an individual, after joining an organization, is similar to the collective behavior of other animals in simpler environments. Stated differently, it can be hypothesized that the new behavior of an individual, after they join an organization, is an emergent property, induced by the critical interaction with the other members of the organization. As we mentioned, such emergent behavior has been observed in a variety of animal groups by naturalists, including, swarms of insects [129], schools of fish [130] and flocks of birds [132] and can be related to the cooperative behavior observed in human society studied by psychologists and sociologists [45].

The organizational behavior, observed in each of these animal groups, demonstrates the need for investigating the across scale dynamics of complex networks. The coupling across scales washes out the dependence on any one scale, and modeling such scale-free behavior, leads to an understanding of how new, large-scale, remote behavior emerges from small-scale local dynamics. These various groups demonstrate collective behavior reminiscent of phase transitions at a critical point (vapor condensing to water, or water freezing to ice) studied by physicists [133], whereas a social scientist might observe a tipping point; each sees a dynamic system undergoing a transformation of phase. Explaining dramatic social transformations in terms of phase transitions was, in fact, done by two physicists, Callen and Shapero in 1974 [131]. They put together the concepts of social imitation and critical behavior, a generation before Gladwell popularized the concept of the tipping point [63].

Despite theoretical, experimental and observational developments, the ability to make scientific predictions of the behavior of complex networks is still in its infancy. The methods of calculus, the mathematics that launched the modern era of physics, have demonstrable limitations, when applied to social and life science. The scientific barriers are a consequence of the fact that social and living systems, in contrast to inert physical systems, are extremely heterogeneous, highly specialized, non-generic, and operate in a non-equilibrium state [45]. For example, the equilibrium state of a living system is death and studying the properties of the latter provides precious little information about the dynamics of the former. In this final chapter it is appropriate to mention the fractional calculus, which was thought for a very long time to be a niche branch of mathematics, might very well be able to span the gap between the inert materials of physics and the living dynamics of social networks.

The fractional calculus was developed in parallel with the classical calculus, and has only recently been shown to be a convenient way to describe the dynamics of complex phenomena. This new approach is particularly important for modeling phenomena characterized by long-term memory and spatial heterogeneity [134 - 136]. The time evolution of fractal processes are determined by fractional differential equations, such as in anomalous diffusion [137, 138], viscoelasticity [139] and turbulent fluid flow [140]. These and many other areas of application are reviewed by West and Grigolini [88, 142].

Of particular interest for this summary is the fact that solution to the simplest fractional rate equation is an inverse power law. The fractional calculus can describe the dynamics of a probability density and thereby give rise to a Pareto distribution; it can describe the dynamics of a fluctuating time series resulting in an inverse power-law correlation function; it can even describe a new kind of non-local force law and generate an inverse power-law response function. In short, each of the 9 effects for generating an inverse power law, discussed in Chapter 4, can be expressed in terms of a dynamic equation, using the fractional calculus, without the often torturous arguments required using the ordinary calculus.

We [45] demonstrated the utility of the fractional calculus in describing the dynamics of the individual elements within a complex network, using the information quantifying that network's global behavior. Using a decision making model, which is based on the assumption of imperfect imitation of one's neighbors [45, 74], we demonstrated that an individual's trajectory response to the collective motion of the network can be described by a linear fractional differential equation. The solution to this fractional equation retains the full influence of the organization on the individual. The comparison of the single individual in isolation to the calculations involving the interaction with the members of the network is depicted in Fig. (**6.1**). The isolated individual is seen to have an exponential probability of changing his/her opinion. The calculation of the individual interacting with the other 9,999 members of the network when the network undergoes a phase transition is shown in red. The dark dashed curve is the solution to the linear fractional rate equation we [45] obtained for the interaction of the individual with the network.

The excellent fit to the network calculation, using a simple fractional rate equation, suggested to me that there was something fundamental here. This result, along with some others not shown, motivated me to present a lecture based on applying the fractional calculus as a fundamental tool for the understanding of complex phenomena. The response to the lecture was so unexpectedly positive, that I decided to write a brief review paper on the same topic [141]. The paper, in turn, generated so much discussion, that I decided to further develop the idea in a book [142]. So if you are interested in where we go from here in the discussion of complexity, and are serious about learning the underlying mathematics, I recommend this book to you.

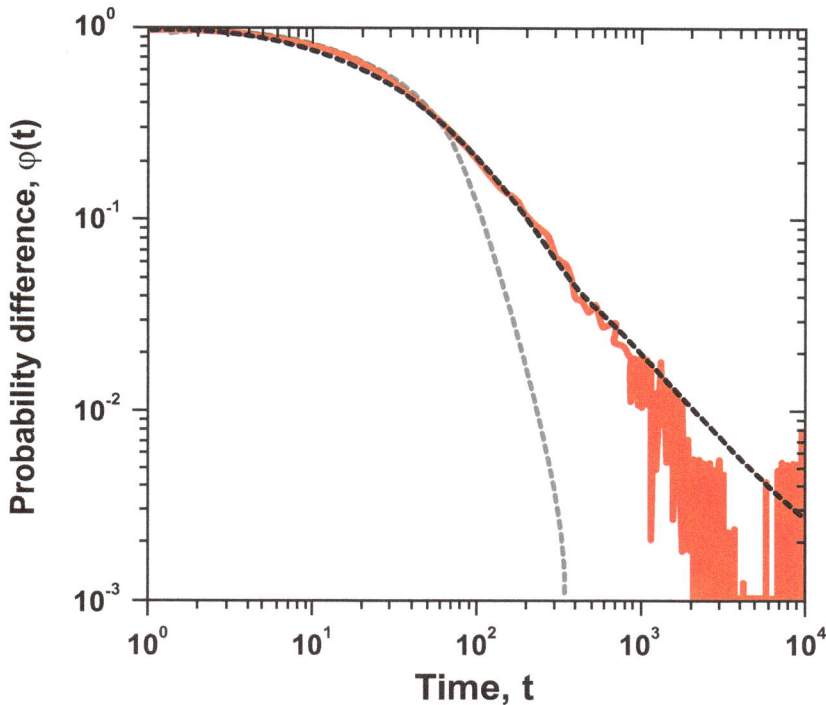

Fig. (6.1). The waiting time distribution before an individual makes a transition is depicted in three calculations. The light dashed curve is an exponential for the behavior of the individual without contact with the network. The red curve is the result of a highly nonlinear decision making model network calculation with a 10^3 people. The dark dashed curve is the solution to the linear fractional calculus equation. [Adapted from [142] with permission].

6.3. A FINAL WORD

The inverse power-law distribution for a complex network has a number of implications for network fragility and robustness against failures, whether through negligence, chance or attacks. Understanding these properties can guide the design of protection strategies against failure, within communications networks, information networks, as well as power grids. This understanding also allows for the formation of preemptive strikes at critical points in terrorists networks. The distinction between robustness and fragility is crucial in this regard. The term fragile means the network remains robust, as certain of its properties (parameters) are changed, but at a critical parameter value there is catastrophic failure that occurs without warning.

The scaling property of complex networks gives the illusion of stability and makes them less vulnerable to random attacks. This property was first observed in physiology, where the fragility of lung architecture was examined. If the mammalian lung scaled classically it would respond too strongly to random fluctuations in genetics, chemistry and the environment and therefore would be maladaptive. Fractal scaling of the kind also observed for the Internet, provided a relative insensitivity to random fluctuations of all kinds. Consequently, the structure of the mammalian lung is pre-adapted to a complex dynamical environment [46, 84]. Similar arguments were applied by others to the Internet, where it was found that the scale-free structure is robust against random attacks. In this context the 80/20 principle would imply that certain nodes in the network have many more connections than do the majority. Therefore it is one of the majority that is destroyed during a random attack and not one of the 'few'. However, it is also clear that if the few can be identified and attacked selectively, then the network would be subject to catastrophic failure. Thus, the scale-free network is fragile. However some engineering design studies of complex networks suggest ways for increasing robustness and decreasing fragility of their scale-free character.

We now realize the prophetic wisdom of a fourteenth century proverb included by Benjamin Franklin in Poor Richard's Almanac in 1758:

> " For want of a nail a shoe was lost,
> For want of a shoe a horse was lost,
> For want of a horse a knight was lost,
> For want of a knight a battle was lost,
> For want of a battle a war was lost,
> And all because of a nail."

In this lyric everything is interconnected and the failure of one component generates a cascade of failures leading to catastrophe. In a modern context this interconnectivity is acknowledged and models are sought to enhance understanding of the implications of individual failure. The interdependency is used to advantage through an integration of man and machine.

Managing networks of increasing complexity is one of the main challenges for decision scientists in the twenty-first century. The national electrical power grids, the global computer networks, and the international software that controls them grow increasingly complex, their fragility to bugs and security flaws become increasingly problematic. It is natural to be concerned over the consequences this growing complexity has on society's ability to manage these networks. A major challenge for network science that must be met over the next decade is the development of techniques by which information is transmitted in a secure fashion to the appropriate level in any organizational structure, whether corporate, government or military. The issues regarding the interrelations of data, information and knowledge, are also tied to communication capabilities, which are distinguished by the highly variable spatial and time links of the social environment.

It is important that we understand the reasons why present day mathematics/science has not described truly complex networks. What is necessary is a new way of thinking in order to understand networks that lack analytic representation, that require scaling and blends regularity and randomness. Some of the mathematical/scientific areas, where these seeds have taken root, are fractal and non-stationary statistics, fractal geometry, nonlinear control theory, and the fractional calculus. However, the application of these mathematical tools to complex networks is still an active area of research, consuming the time and attention of hundreds of scientists around the world.

We conclude that complex networks require a new kind of mathematics for their understanding, and this new mathematics, in turn, entails a new way of thinking. It was disconcerting to the physics community that determinism does not imply predictability, and just because an investigator knows the equations of motion, does not mean the future can be predicted. This was the lesson of nonlinear dynamics and chaos. These nonlinear systems have solutions, whose sensitivity to initial conditions is so great that it does not make practical sense to talk of predicting the final state of the system. This insight has been used to explain our failure, using traditional mathematical models, to describe turbulence, classes of inhomo-geneous chemical reactions, physiological phenomena, the mechanical properties of every day materials such as clay and rubber, and so on, through a host of other complex networks.

Many phenomena are singular and others do not possess a fundamental scale. For example, the terrains on which battles are fought, have historically consisted of sand, stones, rocks, boulders, hills, and mountains, with or without the foliage of grass, reeds, leaves and trees. The mix of these elements on a terrain is such that no one scale characterizes the environment. Consequently, traditional attempts to model the physical battlespace require multi-scale dynamical elements. On the other hand, fractals provide the geometry for such natural settings and suggest new tools for finding man-made objects with sharp corners in such contexts, such as tanks and other weapons platforms. The dynamics of fractals are provided by the fractional calculus [142].

Scaling and fractals are ubiquitous in the world of the complex. Phenomena manifest scaling once their behavior has exceeded a certain richness threshold. Dynamic, economic, social, physical and language networks, all scale, at least in part, and therefore do not lend themselves to traditional dynamics. In fact, the connectedness of networks determines its security against attack. Different architectures for the connections to sites on a network have been shown using various models to yield different levels of robustness. Those networks whose site connections scale, the connectivity is an inverse power law, are the most robust against outside attack. Inverse power laws, in biological networks is indicative of evolutionary advantage [46]; it is feasible that the same is true in complex physical and social networks. Understanding the nature of this advantage guides the design of wireless communication networks, distributed fire control networks, and other such networks within the Army.

A final example of the need for a new way of thinking is demonstrated through examining how pervasive the notion of linearity is in our understanding of the real world. Consider the simplest of control systems, that of aiming a rifle at a moving target. When the aim of the weapon is off the desired target trajectory by a given amount, a negative feedback to the input in direct proportion to the amount of error, corrects the aim and reduces the error to zero. The leading of a moving target is familiar to anyone who has gone duck hunting. Such simple tracking systems were developed under the name cybernetics by the mathematician Wiener in the 1940s and have been extended and developed extensively over the last half century. What has not changed is the assumed linear nature of the error and the error adjustment that is made.

In large-scale complex networks things are different from linear tracking. Many subnetworks have access to different sources of information and must autonomously make their own local decisions, while still cooperating with other subnetworks to achieve a common network-wide goal. Each subnetwork operates with limited knowledge of the overall network structure, which may dynamically reorganize itself due to changing requirements, environmental fluctuations or failure in other subnetworks. Controls are required to be robust with respect to uncertainty and to adapt to fast, as well as, to slow changes in the network dynamics. These considerations make the modern complex networks more like a living network than the mechanical tracking of a gun.

Casti [143] proposes a one-to-one matchup between the problems that control engineers consider to be significant and the problems of life. This match up demands a reformulation of the standard modeling paradigm of control theory. He points out that what is required is extending the conventional input/output and state-variable frameworks upon which modern control theory rests to take explicitly into account the processes of self-repair and replication that characterizes all known living things. Here again the fractional calculus enters through fractional control.

I believe that every author searchers for that one sentence or phrase with which to end their story and that will reverberate in the mind of the reader, such as:

> "It is a far, far better thing that I do, than I have ever done; it is a far, far better rest that I go to than I have ever known."

The sentence or phrase, when it is successful, becomes so memorable that a reference to the original work is not necessary. My modest effort at such an ending is to observe:

"Tomorrow's science will move beyond our arbitrary disciplines and directly address the natural and man-made complexity of the world."

Perhaps I set the bar too high.

6.4. EPILOGUE

As we move between scientific disciplines what is recognized as complexity changes in significant ways. Even with these differences, however, a common characteristic that stands out is the balance between regularity and uncertainty. The significance of this observation cannot be overstated. Those that are whetted to deterministic theories have no more of the scientific truth than those that rely exclusively on the statistical analysis of experimental/observational data, which is to say the have half the answer. The prediction for the outcome of an experiment involving a complex phenomenon will be neither exact, nor have a bell-shaped distribution. The influence of multiple scales, memory and intermittency will deform and distort the prediction made using the traditional tools of science.

Rather than rehashing the content of the book here, let me propose that the effort predicts the future direction science will take, due its continuing attempt to understand complexity. Most predictions of the future of science, technology and their impact on society are an embarrassment to those that made them. This is due to the many ways such predictions can and do fail. So what are the chances of my prediction concerning the future direction of science being right? I do not know the answer to that question, but I can make some conjectures.

Setting aside the fact that scientific discoveries and the new tools that make them possible are completely unpredictable, we briefly focus on technology forecasting. We take this restricted approach because the successes and failures in technology are much better known than the successes and failures in science. The first mode of futurology failure, in order of frequency of occurrence, is assigning too much significance to a technology, because it is attractive and captures the imagination. For example, backpack rockets for the individual soldier have been predicted since the turn of the last century, but with little actual success. The second mode by which forecasts fail, is omission. In this mode one does not properly assess the importance of a nascent technology. For example, almost no one anticipated the impact of the personal computer on society. The third and final mode of failure is guessing the development of the wrong branch of alternative technologies, as in the case of the marketplace choosing VHS over the beta recorder. It certainly was not technical superiority that determined the adoption of the VHS, but some other aspect of the market that is not yet understood. But the point became moot when VHS in its turn was replaced by the DVD.

Of course we are not interested in technological forecasting in itself. It is merely a stepping stone to our main concern, which is forecasting the efficacy of science in the lives of individuals and how they make decisions. I will proceed differently from the above for predicting the future of science and present a list of prediction made by some of the premier scientists of the nineteenth century. These public quotes indicate, without additional commentary, just how wrong even great scientists can be in predicting the future of what they do:

> "When I began my physical studies [in Munich in 1874] and sought advice from my venerable teacher Philipp von Jolly... he portrayed to me physics as a highly developed, almost fully matured science... Possibly in one or another nook there would perhaps be a dust particle or a small bubble to be examined and classified, but the system as a whole stood there fairly secured, and theoretical physics approached visibly that degree of perfection which, for example, geometry has had already for centuries. - from a 1924 lecture by Max Planck (Sci. Am, Feb 1996 p.10)."

> "Sometimes I really regret that I did not live in those times when there was still so much that was new; to be sure enough much is yet unknown, but I do not think that it will be possible to discover anything easily nowadays that would lead us to revise our entire outlook as radically as was possible in the days when telescopes and microscopes were still new.- Heinrich Hertz as a physics student 1884."

> "We are probably nearing the limit of all we can know about astronomy. - Simon Newcomb, early American astronomer 1888."

> "The more important fundamental laws and facts of physical science have

all been discovered, and these are now so firmly established that the possibility of their ever being supplanted in consequence of new discoveries is exceedingly remote.... Our future discoveries must be looked for in the sixth place of decimals. - Albert. A. Michelson, speech at the dedication of Ryerson Physics Lab, U. of Chicago 1894."

"There is nothing new to be discovered in physics now. All that remains is more and more precise measurement- Lord Kelvin 1900."

Given how much the assessment of science, made by these otherwise intellectual giants, missed the mark, I have every confidence that today's science is quite able to expand in ways which I cannot imagine. Each of these great men made the same mistake, they each focused on the known and knowable, and not on the unknown. This book has been about science coming to grips with the complexity of the world, outside the laboratory, where things are barely measurable much less knowable. The explicit theme has been how science has changed in order to understand complexity. The implicit theme has been how our intellectual maps have been modified to accommodate this new kind of understanding.

In the social sciences the formation of consensus for a candidate in an election is separate and distinct from the agreement reached by aggregating the opinion of a large group of independent individuals. Recent insight into social behavior has been gained through the notion of a tipping point, which is the social equivalent of a physical phase transition. The nonlinear interaction between individuals in a social network changes from short-range interactions between friends and family to long-range interactions within town and country. It is possible, using this newly developed complexity formalism, to quantify social phase transitions, including the transition from a peaceful demonstration to a destructive mob, due to the influence of a committed minority.

In ecology we find that time is not the absolute featureless parameter used by Newton, nor is it the velocity-dependent variable introduced by Einstein. Time is a subjective quantity that can be related to the metabolism of an organism. In this way time passes more slowly for the whale and elephant than it does for the shrew and humming bird. This is more than a poetic observation about natural science, it is a quantified determination of how the time experienced by animals is related to their size through an allometry relation. In a related way, the subjective time experienced by the human brain is distinct from chronological time and determines how the brain processes information. This same dichotomy of chronology explains the adaptive change of humans to corporate life.

Let me end these remarks by reiterating the *Principle of Complexity Management*, whereby a system with greater information, but perhaps lesser energy, can dominate a system with lesser information, but greater energy. The principle is a recently proven generalization of an observation made by the mathematician Norbert Wiener in 1948. It has been used to explain habituation, the phenomena of adapting to simple ongoing stimuli that do not have survival value. The principle implies that the information transfer between a lecturer and her audience is maximum, when the lecturer is able to stimulate the audience, with a level of complexity comparable to the listener's brain. This does not just mean the intellectual content of the lecture, but all that goes with it; including dramatic flourishes, gesticulation, changes in intonation patterns. Or stated the other way around, we most readily absorb (manage) information, when the complexity of that information is comparable to the complexity of cognitive processing. In this regard our cognitive map acts like a filter, letting through what matches and blocking, or ultimately suppressing, what does not. Consequently, the more complex the cognitive map, the fewer restrictions there are on what we allow ourselves to understand.

Finally, the more we understand about the world the better we can control our environment. Understanding does not guarantee control, but not understanding guarantees its loss.

References

[1]   R.W. Fogel, "Historiograph and Retrospctive Econometrics", In: B.J. West, Ed., *Mathematical Models as a Tool for the Social Sciences*. Gordon and Breach: New York, 1980, pp. 1-12.

[2]   E.W. Montroll, *On the dynamics and evolution of some sociotechnical systems,* vol. 16. Bull. Am. Math. Soc., 1987, p. 1.

[3]   A. Smith, *A Inquiry into the Nature and Causes of the Wealth of Nations.* W. Strahan and T. Cadell: London, 1776.
        [http://dx.doi.org/10.1093/oseo/instance.00043218]

[4]   K. Marx, and F. Engels, *The Communist Manifesto, English translation from German, 1848.* Penguin Group: New York, 1998.

[5]   T. Carlyle, *On Heroes, Hero Worship, and the Heroes of History.* James Fraser: London, 1841.

[6]   N. Wiener, *Invention: The care and feeding of ideas, manuscript 1954.* The MIT Press: Cambridge, MA, 1993.

[7]   E.W. Montroll, and W.W. Badger, *Introduction to the Quantitative Aspects of Social Phenomena.* Gordon and Breach: New York, 1974.

[8]   T.R. Malthus, *Essay on The Principle of Population As It Aff ects the Future Improvement of Society.* J. Johnson: London, 1798.

[9]   K.E. Boulding, *Collected Papers,* vol. 2. Colorado Associated Univeristy Press, 1971, p. 137.

[10]   P.F. Verhulst, "Mathematical Researches into the Law of Population Growth Increase", *Nouveaux Memoirs de l'Academie Royale des Sciences et Belles-Lettres de Bruxelles,* vol. 18, pp. 1-45, 1845.

[11]   R. Pearl, and L.J. Reed, "On the rate of growth of the population of the United States since 1790 and its mathematical representation", *Proc. Natl. Acad. Sci. USA,* vol. 6, no. 6, pp. 275-288, 1920.
        [http://dx.doi.org/10.1073/pnas.6.6.275] [PMID: 16576496]

[12]   D.H. Meadows, D.L. Meadows, J. Randers, and W.W. Behreus, *The Limts of Growth.* Universe Books: New York, 1972.

[13]   R.M. May, "Simple mathematical models with very complicated dynamics", *Nature,* vol. 261, no. 5560, pp. 459-467, 1976.
        [http://dx.doi.org/10.1038/261459a0] [PMID: 934280]

[14]   J. Davidson, "On the ecology of the growth of the seep population in South Australia", *Trans. R. Soc. S. Afr,* vol. 62, pp. 141-148, 1938.

[15]   T. Modis, and A. Debecker, "Chaoslike states can be expected before and after logistic growth", *Technol. Forecast. Soc. Change,* vol. 41, pp. 111-120, 1992.
        [http://dx.doi.org/10.1016/0040-1625(92)90058-2]

[16]   T. Modis, *Predictions 10 Years Later.* Growth Dynamics: Switzerland, 2002.

[17]   T.C. Fisher, and R.H. Pry, "A simple substitution model of technological change", *Technol. Forecast. Soc. Change,* vol. 3, pp. 75-88, 1971.
        [http://dx.doi.org/10.1016/S0040-1625(71)80005-7]

[18]   L.K. Vanston, and R.L. Hodges, "Technology forecasting for telecommu-nications", *Telekronikk,* vol. 4, pp. 32-42, 2004.

[19]   C.M. Christensen, *The innovator's dilemma: when new technologies cause great firms to fail.* Harvard Business Press: Cambridge, 1997.

[20]   D.J. de Sola Price, *Little Science. Big Science*: New York, 1967.

[21]   T. Modis, "Fractal aspects of natural growth", *Technol. Forecast. Soc. Change,* vol. 47, pp. 63-73, 1994.
        [http://dx.doi.org/10.1016/0040-1625(94)90040-X]

[22]   S. Arbesman, "Quantifying the ease of scientific discovery", *Scientometrics,* vol. 86, no. 2, pp. 245-250, 2011.
        [http://dx.doi.org/10.1007/s11192-010-0232-6] [PMID: 22328796]

[23]   C.F. Gauss, "Theoria motus corporum coelestrium, Hamburg,1809", In: *Dover Eng.Trans.* Theory of Motion of Heavenly Bodies Moving about the Sun in Conic Sections : New York, 1963.

[24]   R. Adrian, "Research concerning the probabilities of the errors which hap-pen in making observations, etc", *The Analyst; or Mathematical Museum,* vol. 1, pp. 93-109, 1809.

[25]   H. Poincaré, *The Foundations of Science.* 1st ed The Science Press: New York, 1929.

[26]   P. Cootner, Ed., *The Random Character of the Stock Market.* MIT press: Cambridge, MA, 1964.

[27]   B.G. Malkiel, *A Random Walk Down Wall Street.* W.W. Norton: New York, 1975. last revision, 2012.

[28]   L. Bachelier, *Annales Scientifiques de l'Ecole Normale Supérieure,* no. 1017, suppl. 3, 1900.

[29]   A. Einstein, *Investigations of the Theory of Brownian Movement.* Dover: New York, 1956.

[30]   J. Perrin, *Brownian Movement and Molecular Reality.* Taylor and Francis: London, 1910;. Alean: Los Atomes, Paris, 1913

[31]   R. Brown, "A brief account of microscopical observations made in the months of June, July and August, 1827, on the particles contained in the pollen of plants; and on the general existence of active molecules in organic and inorganic bodies", *Philos. Mag,* vol. 4, pp. 161-173, 1828.

[32]   J. Ingenhousz, *Dictionary of Scientific Biology.* Scribners, New York, 1973, p. 11.

[33]   F. Galton, "The geometric mean, in vital and social statistics", *Proc. R. Soc. Lond,* vol. 29, pp. 365-367, 1879.
        [http://dx.doi.org/10.1098/rspl.1879.0060]

[34]   F. Galton, *Natural Inheritance.* Richard Clay & Sons: London, 1886.

[35]   A. Quetelet, *Sur 'homme et le developpement des ses faculties, ou essai de physique sociale,* vol. 2. Brussels, 1835.

[36]   T.M. Porter, *The Rise of Statistical Thinking 1820-1900.* Princeton University Press: Princeton, New Jersey, 1986.

[37]   W. Newmarch, "Some observations on the present position of statistical inquiry with suggestion for improving the organization and efficiency of the international statistical congress", *JRSS,* vol. 33, pp. 362-369, 1860.

[38]   S.S. Stevens, "Cross-modality validation of subjective scales for loudness, vibration, and electric shock", *J. Exp. Psychol,* vol. 57, no. 4, pp. 201-209, 1959.
        [http://dx.doi.org/10.1037/h0048957] [PMID: 13641593]

[39]   F.S. Roberts, *Measurement Theory, Encyclopedia of Mathematics and Its Applications.* Addison-Wesley: Reading, MA, 1979.

[40]   J.P. Richter, Ed., The Notebooks of Leonardo da Vinci,

[41]   K. Schmidt-Nielsen, *Scaling; Why is Animal Size So Important.* Cam-bridge University Press: Cambridge, 1984.
        [http://dx.doi.org/10.1017/CBO9781139167826]

[42]   R.E. Chandler, R. Herman, and E.W. Montroll, "Traffic Dynamics: Studies in Car Following", *Oper. Res,* vol. 6, no. 165, 1958.

[43]   S.P. Murray, "Turbulent diffusion of oil spill in the ocean", *Lumnology and Oceanography,* vol. 17, p. 651, 1972.

[44]   J.S. Mill, *A System of Logic, John W.* Parker: London, 1843.

[45]   B.J. West, M. Turalska, and P. Grigolini, *Network of Echoes, Imitation, Innovation and Invisible Leaders.* Springer: Berlin, 2014.

[46]   B.J. West, *Where Medicine Went Wrong; rediscovering the path to com-plexity.* World Scientific: Singapore, 2006.

[47]   S.J. Gould, *Full House.* Harmony Books: New York, 1996.
        [http://dx.doi.org/10.4159/harvard.9780674063396]

[48]   V. Pareto, *Les Systèmes Socialistes,* vol. 2. Paris, 1902. 2nd Edition,1926

[49]   V. Pareto, *Cours d'Economie Politique.* Lausanne, Paris, 1897.

[50]   L. David Roper, *Income Distribution in the United States: A Quantitative Study,* 2007. http://arts.bev.net/RoperLDavid

[51]   B.J. West, and N. Scafetta, *Disrupted Networks: from physics to climate change.* World Scientific: New Jersey, 2010.

[52]   H.M. Gupta, J.R. Campanha, and F.R. Chavorette, "Power-law distribution in high school education: effect of economical, teaching and study conditions", *arXir.0301523v1,* 2003.

[53]   F. Omori, "On the aftershocks of earthquakes", *J. College Sci Imperial Univ. Tokyo,* vol. 7, pp. 111-200, 1894.

[54]   K. Christensen, L. Danon, T. Scanlon, and P. Bak, "Unified scaling law for earthquakes", *Proc. Natl. Acad. Sci. USA,* vol. 99, suppl. Suppl. 1, pp. 2509-2513, 2002.
        [http://dx.doi.org/10.1073/pnas.012581099] [PMID: 11875203]

[55]   P. Hines, J. Apt, and S. Talukdar, "Trends in the istory of large blackouts in the United States", *Energy Policy,* vol. 37, pp. 5249-5259, 2009.
        [http://dx.doi.org/10.1016/j.enpol.2009.07.049]

[56]   B.J. West, and T.R. Clancy, "Flash Crashes, Bursts, and Black Swans: Parallels Between Financial Marketsand Healthcare Systems", *JONA,* vol. 40, pp. 1-4, 2010.

[http://dx.doi.org/10.1097/NNA.0b013e3181f88a8b]

[57]   B.B. Mandelbrot, and R.L. Hudson, *The (Mis)Behavior of Markets: A Fractal View of Risk, Ruin and Reward.* Profile: London, 2004.

[58]   N.N. Taleb, *The Black Swan.* Random House: New York, 2007.

[59]   A.L. Barabasí, *Bursts: The Hidden Patterns Behind Everything We Do.* Penguin Publishing: New York, NY, 2010.

[60]   R.N. Mantegna, and H.E. Stanley, *An Introduction to Econophysics.* Cambridge University Press: UK, 2000.

[61]   B.B. Mandelbrot, *The Fractal Geometry of Nature.* Freeman: San Francisco, CA, 1984.

[62]   L.F. Richardson, "Statistics of Deadly Quarrels", In: J.R. Newman, Ed., *The World of Mathmatics,* vol. 2. Simon and Schuster: New York, 1956.

[63]   M. Gladwell, *Tipping Point,.* Back Bay Books. Little, Brown and Co.: New York, 2000.

[64]   A. Budgor, and B.J. West, "Natural forces and extreme events", In: B.J. West, Ed., *Mathematical Models as a Tool for the Social Sciences.* Gordon and Breach: New York, 1980.

[65]   *Venture Science Fiction Magazine,* 66/2 March,

[66]   A.N. Awan, "Virtual Jihadist media: Function, legitimacy, and radicalising efficacy", *Eur. J. Cult. Stud,* vol. 10, pp. 389-408, 2007.
[http://dx.doi.org/10.1177/1367549407079713]

[67]   Available at: http://en.wikipedia.org/wiki/1%25_rule. (Internet_culture)

[68]   R. Koch, *The 80/20 Principle: The Secret of Achieving More with Less.* Nicholas Brealey Publishing, 1998.

[69]   Norbert Wiener: Collected Works, "Time, Communication and the Nervous System", In: *Eur. J. Cult. Stud,* vol. IV. Cambridge, MA: MIT Press, 20071985, pp. 389-408.

[70]   G. Aquino, M. Bologna, P. Grigolini, and B.J. West, "Beyond the death of linear response theory: criticality of the 1/-noise condition", *Phys. Rev. Lett,* vol. 105, pp. 040601-1, 2010.
[http://dx.doi.org/10.1103/PhysRevLett.105.040601] [PMID: 20867831]

[71]   B.J. West, E. Gelveston, and P. Grigolini, "Maximizing information ex-change between complex networks", *Phys. Rep,* vol. 468, no. 1, 2008.

[72]   D.J. Watts, and S.H. Strogatz, "Collective dynamics of 'small-world' networks", *Nature,* vol. 393, no. 6684, pp. 440-442, 1998.
[http://dx.doi.org/10.1038/30918] [PMID: 9623998]

[73]   T.G. Lewis, *Book of Extremes, Why the 21st Century Isn't Like the 20th Century.* Springer: New York, 2014.

[74]   M. Turalska, M. Lukovic, B.J West, and P. Grigolini, "Complexity and synchronization", *Phys. Rev. E Stat. Nonlin. Soft Matter Phys,* vol. 80, no. 021110, 2009.

[75]   M. Turalska, B.J West, and P. Grigolini, "Temporal complexity of the order parameter at the phase transition", *Phys. Rev. E Stat. Nonlin. Soft Matter Phys,* vol. 83, no. 061142, 2011.

[76]   M. Turalska, B.J. West, and P. Grigolini, "Role of committed minorities in times of crisis ", *Sci. Rept,* vol. 3, no. 1, 2013.
        [http://dx.doi.org/10.1038/srep01371]

[77]   F. Vanni, M. Luković, and P. Grigolini, "Criticality and transmission of information in a swarm of cooperative units", *Phys. Rev. Lett,* vol. 107, no. 7, p. 078103, 2011.
        [http://dx.doi.org/10.1103/PhysRevLett.107.078103] [PMID: 21902433]

[78]   H. Haken, "Cooperative phenomena in systems far from thermal equilibrium and in nonphysical systems", *Rev. Mod. Phys,* vol. 47, pp. 67-121, 1975.
        [http://dx.doi.org/10.1103/RevModPhys.47.67]

[79]   S. Galam, and F. Jacobs, "Effects of inflexible minorities", *Physica A,* vol. 321, pp. 366-376, 2007.
        [http://dx.doi.org/10.1016/j.physa.2007.03.034]

[80]   R.F. Voss, and J. Clarke, "1/f Noise from Systems in Thermal Equilibrium", *Phys. Rev. Lett,* vol. 36, pp. 42-45, 1976.
        [http://dx.doi.org/10.1103/PhysRevLett.36.42]

[81]   R.F. Voss, and J. Clarke, "'1/f noise in music and speech", *Nature,* vol. 258, pp. 317-318, 1975.
        [http://dx.doi.org/10.1038/258317a0]

[82]   R.F. Voss, and J. Clarke, ""1/f noise" in music: Music from 1/f noise", *J. Acoust. Soc. Am,* vol. 63, pp. 258-263, 1978.
        [http://dx.doi.org/10.1121/1.381721]

[83]   T. Musha, P.H.E Meijer, R.D. Mountain, and R.J Soulen , "1/f fluctuations in biological systems", *Sixth International Symposium on Noise in Physical Systems ,* 1981
        [http://dx.doi.org/10.1109/IEMBS.1997.756890]

[84]   B.J. West, and W. Deering, The Lure of Modern Science, *Fractal Thinking.* World Scientific: Singapore, 1995.

[85]   Y. Yu, R. Romero, and T.S. Lee, "Preference of Sensory Neural Codying for 1/f signals", *Phys. Rev. Lett,* vol. 94, no. 108103, 2005.

[86]   P. Andriani, and W. McKelvey, "From Gaussian to Paretian thinking: causes and implications of power laws in organizations", *Organ. Sci,* vol. 20, pp. 1214-1223, 2009.
        [http://dx.doi.org/10.1287/orsc.1090.0481]

[87]   H.A. Simon, "The Architecture of Complexity", *Proc. Am. Philos. Soc,* vol. 106, pp. 467-482, 1962.

[88]   B.J. West, and P. Grigolini, *Complex Webs: Anticitpating the Improbable.* Cambridge University Press: Cambridge, UK, 2012.

[89]   E.W. Montroll, and M.F. Shlesinger, "Maximum entropy formalism, frac-tals, scaling phenomenon and 1/f noise: a tale of tails", *J. Stat. Phys,* vol. 32, pp. 209-230, 1983.
        [http://dx.doi.org/10.1007/BF01012708]

[90]   N. Scafetta, S. Picozzi, and B.J. West, "An out-of-equilibrium model of the distribution of wealth", *Quant. Finance,* vol. 4, pp. 353-364, 2004.
        [http://dx.doi.org/10.1088/1469-7688/4/3/010]

[91]   W.W. Calder III, *Size, Function and Life History.* Harvard University Press: Cambridge, MA, 1984.

[92]   Sarrus et Rameaux, "Rapport sur un memoire adresse a L'Academie Royle de Medcine. Commissaires Robiquet et Thillarye, rapporteurs", *Bull. Acad. Roy. Med. (Paris),* vol. 3, pp. 1094-1000, 1838-39.

[93]   M. Rubner, "Ueber den Einfluss der Körpergrösse auf Stoffund Kraftwech-sel", *Z. Biol,* vol. 18, pp. 353-358, 1883.

[94]   M. Kleiber, "Body size and metabolism", *Hilgarida,* vol. 6, pp. 315-353, 1932. [http://dx.doi.org/10.3733/hilg.v06n11p315]

[95]   A.M. Hemmingsen, "The relation of standard (basal) energy metabolism to total fresh weight of living organisms", In: *Rep. Steno. Mem. Hosp,* vol. Vol. 4. Copen-hagen, 1950, pp. 1-58.

[96]   D. West, and B.J. West, "On Allometry", *Int. J. Mod. Phys. B,* vol. 26, pp. 1230013-1, 2012. [http://dx.doi.org/10.1142/S0217979212300101]

[97]   L.R. Taylor, "Aggregation, variance and the mean", *Nature,* vol. 189, pp. 732-735, 1961. [http://dx.doi.org/10.1038/189732a0]

[98]   L.R. Taylor, and R.A. Taylor, "Aggregation, migration and population mechanics", *Nature,* vol. 265, no. 5593, pp. 415-421, 1977. [http://dx.doi.org/10.1038/265415a0] [PMID: 834291]

[99]   A.L. Goldberger, and B.J. West, "Fractals: a contemporary mathematical concept with applications to physiology and medicine", *Yale J. Biol. Med,* vol. 60, pp. 104-119, 1987.

[100] R.K. Merton, "The Matthew Effect in Science", *Science,* vol. 159, pp. 56-63, 1968. [http://dx.doi.org/10.1126/science.159.3810.56]

[101] H. Zuckerman, *Scientific Elite: Nobel Laureates in the United States.* Transaction Publishers: New York, 1995.

[102] D.J. Watts, *Small Worlds: The Dynamics of Networks between Order and Randomness.* Princeton University Press: New Jersey, 1999.

[103] A-L. Barabási, and R. Albert, "Emergence of scaling in random networks", *Science,* vol. 286, no. 5439, pp. 509-512, 1999. [http://dx.doi.org/10.1126/science.286.5439.509] [PMID: 10521342]

[104] G.U. Yule, "A mathematical theory of evolution based on the conclusions of Dr. J. C. Willis", *Philos. Trans. R. Soc. Lond., B,* vol. 213, pp. 21-87, 1925. [http://dx.doi.org/10.1098/rstb.1925.0002]

[105] J.C. Willis, *Age and area: a study in geopraphical distribution and origin of species.* Cambridge University Press, 1922. [http://dx.doi.org/10.5962/bhl.title.70451]

[106] K. Lindenberg, and B.J. West, *The Nonequilibrium Statistical Mechanics of Open and Closed Systems.* VCH: New York, 1990.

[107] M.O. Vlad, and B. Schonfisch, "Mass action law versus contagion dynam-ics. A mean-field statistical approach with application to the theory of epidemics", *J. Phys. Math. Gen,* vol. 29, pp. 4895-4913, 1996. [http://dx.doi.org/10.1088/0305-4470/29/16/015]

[108] S. Kauffman, *At Home in the Universe: The Search for Laws of Self-Organization and Complexity.* Oxford University Press: Oxford, UK, 1995.

[109] B.J. West, and P. Grigolini, "Habituation and 1/f noise", *Physica A,* vol. 389, no. 5706, 2010.

[110] M.V. Simkin, and V.P. Roychowdhury, "Re-inventing Willis", *Phys. Reports,* vol. 502, pp. 1-35, 2011.

[111] P. Allegrini, D. Menicucci, R. Bedini, L. Fronzoni, A. Gemignani, P. Grigolini, B. J. West, and P. Paradisi, "Spontaneous brain activity as a source of ideal 1/f noise", *Phys. Rev. E,* vol. 80, p. 061914, 2009.

[112] D. Nozaki, D.J. Mar, P. Grigg, and J.J. Collins, "Effects of Colored Noise on Stochastic Resonance in Sensory Neurons", *Phys. Rev. Lett.,* vol. 82, pp. 2402-2404, 1999.

[113] S. Vrobal, *Fractal Time: Why a Watched Kettle Never Boils, Studies of Nonlinear Phenomena in Life Science ,* vol. 14. World Scientific: Singapore, 2010.

[114] T. Mann, *Der Zauberberg, Fischer, Frankfurt* 1984 translated in *Fractal Time by Susie Vrobel,*

[115] H. Ebbinghaus, *Über das Gedächtnis (1885); translated to English as Memory. A Contribution to Experimental Psychology, Dover Publ.* Mine-ola: New York, 1964.

[116] G.R. Loftus, "Evaluating forgetting curves", J. Exp. Psych.: Learning", *Mem. Cognit,* vol. 11, pp. 397-406, 1985.
[http://dx.doi.org/10.1037/0278-7393.11.2.397]

[117] B.J. West, and P. Grigolini, "Chipping away at memory", *Biol. Cybern,* vol. 103, no. 2, pp. 167-174, 2010.
[http://dx.doi.org/10.1007/s00422-010-0394-6] [PMID: 20517616]

[118] L. Zemanova, C. Zhou, and J. Kurths, "Structural and functional clusters of complex brain networks", *Physica D,* vol. 224, pp. 202-212, 2006.
[http://dx.doi.org/10.1016/j.physd.2006.09.008]

[119] M.R. Pagani, K. Oishi, B.D. Gelb, and Y. Zhong, "The phosphatase SHP2 regulates the spacing effect for long-term memory induction", *Cell,* vol. 139, no. 1, pp. 186-198, 2009.
[http://dx.doi.org/10.1016/j.cell.2009.08.033] [PMID: 19804763]

[120] D. Ariely, *Predictably Irrational.* Happer Collins: New York, 2008.

[121] M. Wittmann, and M.P. Paulus, "Decision making, impulsivity and time perception", *Trends Cogn. Sci. (Regul. Ed.),* vol. 12, no. 1, pp. 7-12, 2008.
[http://dx.doi.org/10.1016/j.tics.2007.10.004] [PMID: 18042423]

[122] M. Wittmann, D.S. Leland, and M.P. Paulus, "Time and decision making: differential contribution of the posterior insular cortex and the striatum during a delay discounting task", *Exp. Brain Res,* vol. 179, no. 4, pp. 643-653, 2007.
[http://dx.doi.org/10.1007/s00221-006-0822-y] [PMID: 17216152]

[123] G.T. Fechner, *Elemente der Psychophysik.* Breitkopf und Hartel: Leipzig, 1860.

[124] T. Takahashi, "A comparison of intertemporal choices of oneself versus some else based on Tsallis'statistics", *Physica A,* vol. 385, pp. 637-644, 2007.
[http://dx.doi.org/10.1016/j.physa.2007.07.020]

[125] J. Garstka, and D. Alberts, *Network Centric Operations Conceptual Framework, Version 2.0.* 2004. unpublished

[126] D.S. Alberts, J.J. Garstka, and F.P. Stein, *Network Centric Warfare: Developing and leveraging information superiority.* 2nd ed 2004. CCRP Publication Series, ISBN 1-57906-019-6

[127] N. Scafetta, and B.J. West, "Multiscaling comparative analysis of time series and a discussion on "earthquake conversations" in California", *Phys. Rev. Lett,* vol. 92, no. 13, p. 138501, 2004. [http://dx.doi.org/10.1103/PhysRevLett.92.138501] [PMID: 15089646]

[128] W.H. Whyte, *The Organization Man.* Simon & Schuster: New York, 1956.

[129] C.A. Yates, R. Erban, C. Escudero, I.D. Couzin, J. Buhl, I.G. Kevrekidis, P.K. Maini, and D.J. Sumpter, "Inherent noise can facilitate coherence in collective swarm motion", *Proc. Natl. Acad. Sci. USA,* vol. 106, no. 14, pp. 5464-5469, 2009. [http://dx.doi.org/10.1073/pnas.0811195106] [PMID: 19336580]

[130] Y. Katz, K. Tunstrøm, C.C. Ioannou, C. Huepe, and I.D. Couzin, "Inferring the structure and dynamics of interactions in schooling fish", *Proc. Natl. Acad. Sci. USA,* vol. 108, no. 46, pp. 18720-18725, 2011. [http://dx.doi.org/10.1073/pnas.1107583108] [PMID: 21795604]

[131] E. Callen, and D. Shapero, "A theory of social imitation", *Phys. Today,* vol. 27, no. 23, 1974.

[132] A. Cavagna, A. Cimarelli, I. Giardina, G. Parisi, R. Santagati, F. Ste-fanini, and M. Viale, *Proc. Natl. Acad. Sci. USA,* vol. 107, no. 11865, 2010.

[133] H.E. Stanley, *Introduction to Phase Transition and Critical Phenomena.* Oxford University Press: New York, 1971.

[134] R.L. Magin, *Fractional Calculus in Bioengineering.* Begell House: CT, 2006.

[135] I. Podlubny, *Fractional Diff erential Equations, MATHEMATICS in SCI-ENCE and ENGINEERING,* vol. 198. Academic Press: San Diego, 1999.

[136] B.J. West, M. Bologna, and P. Grigolini, *Physics of Fractal Operators.* Springer: New York, 2003. [http://dx.doi.org/10.1007/978-0-387-21746-8]

[137] R. Metzler, and J. Klafter, "The random walk's guide to anomalous diffu-sion: a fractional dynamics approach", *Phys. Rep,* vol. 339, pp. 1-77, 2000. [http://dx.doi.org/10.1016/S0370-1573(00)00070-3]

[138] V. Seshadri, and B.J. West, "Fractal dimensionality of Lévy processes", *Proc. Natl. Acad. Sci. USA,* vol. 79, no. 14, pp. 4501-4505, 1982. [http://dx.doi.org/10.1073/pnas.79.14.4501] [PMID: 16593212]

[139] W.G. Glöckle, and T.F. Nonnenmacher, "A fractional calculus approach to self-similar protein dynamics", *Biophys. J,* vol. 68, no. 1, pp. 46-53, 1995. [http://dx.doi.org/10.1016/S0006-3495(95)80157-8] [PMID: 7711266]

[140] M.F. Shlesinger, B. J. West, and J. Klafter, "Lévy dynamics of enhanced diffusion: Application to turbulence", *Phys. Rev. Lett,* vol. 58, no. 1100, 1987.

[141] B.J. West, "Fractional calculus view of complexity: A tutorial", *Rev. Mod. Phys,* vol. 86, pp. 1169-1184, 2014.

[http://dx.doi.org/10.1103/RevModPhys.86.1169]

[142]   B.J. West, *Tomorrow's Science; A Fractional Calculus View of Complexity.* CRC: New York, 2015.
[http://dx.doi.org/10.1201/b18911]

[143]   J.L. Casti, ""Biologizing" Control Theory: How to make a control system come alive", *Complexity,* vol. 7, pp. 10-12, 2002.
[http://dx.doi.org/10.1002/cplx.10032]

SUBJECT INDEX